CHICAGO PUBLIC LIBRARY
HAROLD WASHINGTON LIBRARY CENTER

R0016697205

REF
OH   Forbes, Edward,
135   1815-1854.
.F67
1977   The natural history
Cop.1   of the European
         seas

DATE			

REF
QH
135
.F67
1977
Cop.1
Received

FORM 125 M

Business/Science/Technology Division

The Chicago Public Library

FEB 24 1979

© THE BAKER & TAYLOR CO.

THE NATURAL HISTORY
OF THE EUROPEAN SEAS

This is a volume in the Arno Press collection

HISTORY OF ECOLOGY

Advisory Editor
Frank N. Egerton III

Editorial Board
John F. Lussenhop
Robert P. McIntosh

*See last pages of this volume for a
complete list of titles.*

THE NATURAL HISTORY

OF

THE EUROPEAN SEAS

Edw[ard] Forbes
and
Robert Godwin-Austen

ARNO PRESS
A New York Times Company
New York / 1977

Editorial Supervision: LUCILLE MAIORCA

Reprint Edition 1977 by Arno Press Inc.

Reprinted from a copy in
 The University of Illinois Library

HISTORY OF ECOLOGY
ISBN for complete set: 0-405-10369-7
See last pages of this volume for titles.

Manufactured in the United States of America

Publisher's Note: This map has been reproduced in black and white in this edition.

Library of Congress Cataloging in Publication Data

Forbes, Edward, 1815-1854.
  The natural history of the European seas.

  (History of ecology)
  Reprint of the 1859 ed. published by J. Van Voorst, London, which was issued as v. 2 of Outlines of the natural history of Europe.
  1. Marine biology—Europe. 2. Marine biology.
I. Godwin-Austin, Robert Alfred Cloyne, 1808-1884, joint author. II. Title. III. Series. IV. Series: Outlines of the natural history of Europe ; v. 2.
QH135.F67   1977   591.9'2'13   77-74221
ISBN 0-405-10392-1

18,80

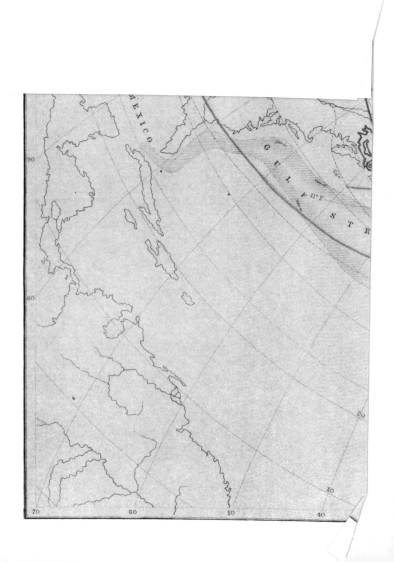

OUTLINES
OF
THE NATURAL HISTORY OF EUROPE.

THE

NATURAL HISTORY

OF

THE EUROPEAN SEAS.

BY THE LATE
PROF. EDW. FORBES, F.R.S.,
ETC.

EDITED AND CONTINUED
BY
ROBERT GODWIN-AUSTEN, F.R.S.

LONDON:
JOHN VAN VOORST, PATERNOSTER ROW.
MDCCCLIX.

LONDON: PRINTED BY WOODFALL AND KINDER,
ANGEL COURT, SKINNER STREET.

CONTENTS.

CHAP.		PAGE
I.	INTRODUCTORY	1
II.	ARCTIC PROVINCE	28
III.	BOREAL PROVINCE	59
IV.	CELTIC PROVINCE	78
V.	LUSITANIAN PROVINCE	106
VI.	MEDITERRANEAN PROVINCE	127
VII.	THE BLACK SEA	200
VIII.	THE CASPIAN SEA	209
IX.	ON THE DISTRIBUTION OF MARINE ANIMALS	217
X.	EARLY HISTORY OF THE EUROPEAN SEAS	248
XI.	CONCLUSION, AND EXPLANATION OF THE MAP	279

PREFACE.

This volume requires a few words of explanation and apology. It is now some years since its Publisher proposed to issue a series of volumes on the "Outlines of the Natural History of Europe." Prof. Henfrey undertook the subject of "The Vegetation of Europe," Prof. Ed. Forbes that of "The Natural History of the European Seas," and he proposed to me that I should write the "Geological History of the European Area."

Prof. Henfrey's volume appeared in 1852, and it was then announced that the second volume of the series would be by Prof. Ed. Forbes, and would appear in 1853. The volume, however, did not make its appearance, nor (by those who were acquainted with the Author) could it be reasonably expected ; at that time much work and many engagements pressed heavily on him. Apart from this, his own special studies, as well as the researches of others, particularly those of his friend Mr. M'Andrew, caused him, I think, to wish for a little delay before he committed himself to any general views as to the History of European Marine Fauna as a whole.

In 1854 Ed. Forbes was elected to the Natural History Chair at Edinburgh, and it was his intention to have finished this work in the course of that winter, so that it should appear in 1855. But it was not to be so ; and that active life, which in the last few years had accomplished so much, which was then proposing so much for the future, and of which the past was the ample pledge how productive the succeeding years would have been made, was brought to its too sudden close.

It was some time after this event that I received a portion of the present volume—as far as page 102, this had been corrected and printed off; the latest portion of the Manuscript which the Author had forwarded, had been set up in type, but not corrected, and has served to bring down Prof. Ed. Forbes' portion as far as to page 126.

It had been my friend's last wish, founded on

too partial an estimate, that I should take charge of such works as he was then engaged on. It was in this way that I edited the "Memoir on the Tertiary Fluvio-Marine Formation of the Isle of Wight."

When, somewhat later, it was proposed that I should undertake the continuation of the "Natural History of the European Seas," I shrank from the difficulty of the task; but I saw that the completed portion of the work was too slight to be issued by itself, and I was unable to find anything by the author which could be added, and it was solely to promote the publication of what competent judges assured me ought not to be withheld, that I undertook to carry out, to the best of my ability, the plan of the author, as that may be gathered from page 16, and some other passages.

For a continuation to be successful, the graft must be better than the original stock. The subject of the present volume partakes, necessarily, of the nature of an enumeration, and it was not an encouraging task to carry on my friend's facile style of natural history narrative. Again, though he repeatedly admits that the present is inseparably connected with the past, yet in reviewing my portion of this joint volume, I feel that I may be charged with having treated the subject too often from a geological stand-point, and that my share may possibly remind some of the Angel of the Sign-painter who had painted Red Lions all his life—and which, in spite of his efforts, looked more like a Lion, than became an Angel, after all.

R. G.-A.

THE NATURAL HISTORY

OF THE

EUROPEAN SEAS.

CHAPTER I.

INTRODUCTORY.

EVERY one knows that the same animals and plants are not found everywhere on the surface of the land, but that they are distributed so as to be gathered together in distinct zoological and botanical provinces, of greater or less extent, according to their degree of limitation by physical conditions, whether features of the earth's outline or climate. Each province is not so entirely distinct from its neighbours as to be exclusively inhabited by creatures peculiar to itself, but shares more or less in the population of those regions which impinge upon its boundaries; so that the line between these zoological and botanical kingdoms, or rather republics, is not sharp and defined, like that which marks the limits of political states, but is softened off and melted, as it were, into the margins of the neighbouring territories; nor, in most cases, is it easy or possible to say where the one terminates and the next begins. In this respect we are reminded of the divisions or classes of animated beings which are not sharply defined sections, but battalions of similar creatures arrayed around distinct types or banners, yet with many irregular troops on the skirts of each, wearing no sufficiently marked uniform, but so attired as now to be claimed by the one, now by the other army.

In this age of volumes, a man had needs offer a good excuse before adding a new book, even though it be a small one, to the heap already accumulated. He should either have something fresh to say, or be able to tell that which is old in a new and pleasanter way. Naturalists and others, whose vocation is the prosecution of science, have an easier task in this respect than their literary brethren. Our volumes may be, and often, from the very nature of their themes, are, comparatively dry and heavy. Yet the adding an ear, or even a grain of wheat to the great granary of human knowledge, whence the brains of future generations are to be nourished, is some small service to the good cause of enlightenment; although we may fold it in unnecessarily many sheets of paper, esteeming it possibly over much because we ourselves have gathered it. How much of the following pages is good grain, and how much husk, it is not for me to judge; but having, when pursuing researches in various parts of the European seas, found the want of some book capable of affording a general view of their natural history, and after a fair amount of personal experience in their exploration, still feeling that the same want must perplex and impede the researches of those who are beginning similar inquiries, I venture on an attempt to fill the blank, with a fair conscience and unabashed. Moreover it is becoming that Britons, whether scientific or unscientific, who boast at all fitting occasions of their aptitude to rule the waves, should know something of the population of their saline empire, especially of those parts of it immediately in contact with their terrestrial domain, and the coasts of the Continent to which our United Kingdom appertains. In the following chapters I have endeavoured to lay before my readers, in plain discourse, and with as few technicalities as possible, the leading features of the several regions of the European seas, to show how they are connected, and how they differ, and to attempt to explain the causes of their peculiarities.

"I do not pretend," wrote Robert Boyle, "to have visited the bottom of the sea; but since none of the naturalists whose writings I have yet met with, have been there any more than I; and it is a great rarity in those cold parts of Europe to meet with any men at all that have had at once the boldness, the occasion, the opportunity, and the skill to penetrate into those concealed and dangerous recesses of nature, much less to make any stay there, I presume it will not be unpleasant, if about a subject of which, though none of those very few naturalists that write anything at all, write otherwise than by hearsay, I recite in this place what I learned by inquiry from those persons, that, among the many navigators and travellers I have had opportunity to converse with, were the likeliest to give me good information about these matters."

Since the days of the illustrious experimental philosopher, naturalists have made great advances in their knowledge of submarine phenomena, and can now speak from their own knowledge, though, for even a good half century after Boyle's censure had been written, they were mainly dependent for their acquaintance with the depths of the ocean on travellers and mariners, whose powers of observation were untrained, and who had no initiation into the elements of natural history. Naturalists have even visited personally the bottom of the sea, for of late years the diving-bell has, in some instances, been used as a means of scientific research, one, however, of limited application; the net and the dredge are the surest means at our command for exploration of the ocean's recesses, especially the last-named instrument, the full use and value of which, however, can scarcely be said to have been understood until within the last twenty years.

The naturalists of yore esteemed the ocean to be a treasury of wonders, and sought therein for monstrosities and organisms contrary to the law of nature, such as they interpreted it. The naturalists of our own time hold equal faith in the wonders of the sea, but seek therein rather for the links of nature's chain than for apparent exceptions. Out of the waves they draw subjects for their most patient and elaborate researches, for the creatures that live beneath the waters exhibit more varied and extraordinary conformations than those that dwell upon the land. Moreover, a great part of them consists of beings in a manner rudimentary—creatures exhibiting the elements of higher creatures, living analyses of higher organized compounds, the first draughts of sketches afterwards finished, the framework, as it were, of many-wheeled machines. By an examination and study of them we get at a clearer conception of the nature of the structures which, in combination, constitute the complicated bodies of vertebrated animals, and in the end are enabled to throw light upon the organization of man himself, learning thereby much concerning the wonderful construction of the microcosm, and at the same time, through our better knowledge of the nature and capabilities of our organization, acquiring a lesser, though more practical gain, in the placing of the science of medicine on a surer and sounder foundation. The day has gone by when a medical student was taught the anatomy and physiology of man with little or no reference to that of inferior beings.

As it is on the land, so is it in the sea; not that, as old philosophers fancied, every terrestrial creature has its double beneath the waves, but that the submarine population is grouped into geographical provinces, which, though well marked in their more central and most developed portions, are merged at their bounds indistinguishably into the edges of neighbouring realms. These submarine provinces have a more or less distinct correspondence with those of the neighbouring lands, though sometimes they differ very considerably from the latter in their extent, since the physical features which may constitute boundaries in the one, may not be sufficiently extended or developed in the other as to impede the spread of peculiar species of animals and plants, or of one or the other only, as the case may be. Marine creatures, owing to their organization and the transporting powers of the element in which they live, are much more capable of diffusion as a whole, than terrestrial; hence we should expect to find the regions into which they are grouped beneath the waves, of much vaster dimensions than those constituted by the geographical assemblages of their terrestrial brethren; and such is, to a great extent, true. Nevertheless, the inequalities of the sea-bed, the modifications of the temperature of the ocean, produced by currents pouring through it like mighty rivers, and leaving with them the climate—sometimes more genial, sometimes more rigorous of the latitudes whence they have derived their source, the intersection of arms or promontories of land, and the more powerful interruptions caused by the great gulfs and abysses of the deep, or by vast and comparatively desert tracts of unprolific sand, which in many places are spread out in extensive shallows, are all powerful influences affecting the diffusion of marine creatures, and determining their distribution within certain and more or less defined limits.

A province, as understood in the following chapters, is an area within which there is evidence of the special manifestations of the Creative Power; that is to say, within which there have been called into being, the originals, or protoplasts, of animals or plants. These may become mixed up with emigrants from other provinces, even exceeding in their numbers the aborigines, so to call them, of the region to which they have migrated. The distinguishing of the aboriginal from the invading population, and the determination of the causes which have produced and directed the invasion, are among the problems which the investigator of the distribution of animated creatures, has to endeavour to solve. When the Fauna or Flora of a province has been thoroughly investigated, the diffusion of the individuals of the characteristic species is found to indicate that the manifestation of the creative energy has not been equal in all parts of the area, but that in some portion of it, and that usually more or less central, the genesis of new beings has been more intensely exerted than elsewhere. Hence, to represent a province diagrammatically, we might colour a nebulous space, in which the intensity of the hue would be exhibited towards the centre, and become fainter and fainter towards the circumference. This feature of zoological and botanical provinces gives rise to the term *centres of creation*, which I and others have applied to them. There may be minor centres within a province. Nowhere do we find a province repeated; that is to say, in none, except one centre of creation do we find the same assemblage of *typical* species; or, in other words, no species has been called forth originally in more areas than one. Similar species, to which the term *representative* is mutually applied, appear in areas distant from each other, but under the influence of similar physical conditions. But every true species presents in its individuals, certain features, *specific characters*, which distinguish it from every other species; as if the Creator had set an exclusive mark or seal, on each living type. Species, the individuals of which are distributed over an unbroken area, exhibit the phenomenon of centrality within themselves, *i.e.*, there is some portion of that area, whence all the individuals of the species appear to have radiated. As from all the facts we know, the relationship of the individuals of any species to each other, exhibits the phenomenon of descent, since every case in which the parentage of an individual or group of similar individuals has been traced, the parent stock has been found similar to it or them, we connect the idea of descent with the definition of a species, and (hypothetically) assume the descent of all the individuals of each species from one original stock, monœcious or diœcious, as the case may be. The term *specific centre* has been used to express that single point upon which each species had its origin, and from which its individuals become diffused. In the course of their diffusion, and during the lapse of time, the species may become extinguished in its original centre, and exist only on some one or several portions of the area over which it became diffused. Groups of the individuals of a single species may thus become isolated, and if they be placed far apart, may present the fallacious aspect of two or more centres for the same species. To get at the causes of such phenomena, we must trace the history of the species backwards in time, and inquire into its connection with the history of geological change. We thus trace the genealogy of the species, and unless there has been any endeavour made to develop its pedigree, and to connect its history in space with its history in time, no man has a right to cite anomalous and isolated cases of distribution, as arguments against the doctrine of specific centres. In studying the geographical distribution of organized beings philosophically, it is absolutely necessary to call in the aid of geology; and the time is not far distant when no reasonable man will venture on that most interesting branch of natural history research without a grounding in geological science.

Provinces, to be understood, must be traced back, like species, to their history and origin in past time. Palæontological research exhibits, beyond question, the phenomenon of provinces in time, as well as provinces in space. Moreover, all our knowledge of organic remains teaches us, that species have a definite existence, and a centralization in geological time as well as in geographical space, and that *no species is repeated in time.* The distribution of the individuals of fossil species also indicates their diffusion from some unique point of origin, and, consequently, goes to support the notion of the connection of these individuals through the relationship of descent, and the derivation of them all from an original protoplast.

The investigation and determination of the provinces of marine life, have as yet been but little pursued, and there is no finer field for discovery in natural history, than that presented by the bed of the ocean, when examined with a view to the defining of its natural subdivisions. The difficulties which attend the inquiry add to the zest of the research; and there is a charm in travelling mentally over the hills and valleys buried inaccessibly beneath their thick atmosphere of brine, unbreathable by mortal lungs, which air-travelling, being an easy possibility, and its results, do not possess. Yet if we be careful never to let our imagination get the better of our judgment, and never to come to a conclusion unless we find that, on strict and logical examination of our reasoning, we have arrived at it through fair means and firm walking, not by leaping over difficulties with closed eyes, we are quite as safe under water as above it, and have as sure footing on the slippery surface of the sea's floor, as on the grassy plain, or rocky mountain. I can speak personally as to the pleasure of such explorations, the more to be esteemed, since in these days there are few countries so entirely new as to warrant the traveller's boast, that he is the first educated man to visit them, and to discover their wonders. But, beneath the waves, there are many dominions yet to be visited, and kingdoms to be discovered; and he who venturously brings up from the abyss enough of their inhabitants to display the physiognomy of the country, will taste that cup of delight, the sweetness of whose draught those only who have made a discovery know. Well do I remember the first day when I saw the dredge hauled up after it had been dragging along the sea-bottom, at a depth of more than one hundred fathoms. Fishing-lines had now and then entangled creatures at as great, and greater depths, but these were few and far between, and only served to whet our curiosity, without affording the information we thirsted for. They were like the few stray bodies of strange red men which tradition reports to have been washed on the shores of the Old World, before the discovery of the New, and which served to indicate the existence of unexplored realms inhabited by unknown races, but not to supply information about their character, habits, and extent. But when a whole dredgeful of living creatures from the unexplored depth appeared, it was as if we had alighted upon a city of the unknown people, and were able, through the numbers and varieties taken, to understand what manner of beings they were. Well do I remember anxiously separating every trace of organic life from the enveloping mud, and gazing with delighted eye on creatures hitherto unknown, or on groups of living shapes, the true habitats of which had never been ascertained before, nor had their aspect, when in the full vigour and beauty of life ever before delighted the eye of a naturalist. And when, at close of day, our active labours over, we counted the bodies of the slain, or curiously watched the proceedings of those whom we had selected as prisoners, and confined in crystal vases, filled with a limited allowance of their native element, our feelings of exultation were as vivid, and surely as pardonable, as the triumphant satisfaction of some old Spanish "Conquisatador," musing over his siege of a wondrous Astlan city, and reckoning the number of painted Indians he had brought to the ground by the prowess of his stalwart arm.

To sit down by the sea-side at the commencement of ebb, and watch the shore gradually uncovered by the retiring waters, is as if a great sheet of hieroglyphics — strange picture-writing — were being unfolded before us. Each line of the rock and strand has its peculiar characters inscribed upon it in living figures, and each figure is a mystery, which, though we may describe the appearance in precise and formal terms, has a meaning in its life and being beyond the wisdom of man to unravel. How many and how curious problems concern the commonest of the sea-snails creeping over the wet sea-weed! In how many points of view may its history be considered! There are its origin and development—the mystery of its generation—the phenomena of its growth—all concerning each apparently insignificant individual; there is the history of the species—the value of its distinctive marks — the features which link it with higher and lower creatures — the reason why it takes its stand where we place it in the scale of creation—the course of its distribution—the causes of its diffusion—its antiquity or novelty—the mystery (deepest of mysteries) of its first appearance— the changes of the outline of continents and of oceans which have taken place since its advent, and their influence on its own wanderings. Some of these questions may be clearly and fairly solved; some of them may be theoretically or hypothetically accounted for; some are beyond all the subtlety of human intellect to unriddle. I cannot revolve in my mind the many queries which the consideration of the most insignificant of organized creatures, whether animal or vegetable, suggests, without feeling that the rejection of a mystery, because it is a mystery, is the most besotted form of human pride.

The sea-board of Europe, exclusive of Iceland, extends through four degrees of latitude, and six of longitude, occupying three sides of an irregular quadroid. The northern, and narrowest, side, lies within the Arctic Circle, is partly included in the Icy Sea, and presents a deeply serrated outline, indented in its centre by the great arm or gulf, known as the White Sea. The western side exhibits all varieties of conformation; in its northernmost and Norwegian portion, it is belted with small islands, and indented with fiords. At the southern termination of Norway we have the tortuous gulfs conducting to the Baltic Sea. The coasts of Denmark and Holland form a tame boundary to the shallow portion of the North Sea, itself originating in the projection northwards of the group of islands of which Great Britain and Ireland are the chief. The deep bend of the Bay of Biscay carries us southward, with a simple outline, to the junction of France and Spain, and to the rocky and partially jagged coasts of Asturias, from whence to the end of Europe, at the Pillars of Hercules, a tame, and but slightly-varied line prevails. The southern side is of great extent and variety, forming as it does, the wavy and irregular margin of the Mediterranean, with its deep arms of the Adriatic and Egean, and continued to make the tamer bounds of the Euxine. A last and isolated portion is that which terminates Europe on the south-east, and constitutes the north-western border of the Caspian Sea.

Along such a range of shore, extending through various climates, from the warm and sunny confines of Africa to the ice-bound cliffs of Nova Zembla and Spitzbergen, we cannot fail to find many and diversified assemblages of animated creatures. The beings who delight in the chilly waters of the Arctic Ocean must be very different from those which revel in the genial seas of the south; whilst the temperate tides that lave our own favoured shores, cherish a submarine population intermediate in character between both. Thus in our progress from north to south we pass through regions or belts exhibiting successive changes in the features of animated nature. It is not so, however, in proceeding from the Straits of Gibraltar to the easternmost recesses of the Mediterranean; passing along the same parallel of latitude throughout, we carry with us, as it were, the creatures who met us at the gates, and when we enter the less pleasant expanse of the Black Sea, we find the differences lie mainly in deficiencies, and not marked by the presence of new creatures. In the inland and isolated Caspian, it is true, we behold strange and peculiar animals, but their presence, as we shall hereafter learn, is rather to be regarded in connection with the past than with the present—as the living witnesses of preadamic ages than as members of the community of creatures characteristic of the epoch in which we live.

This extensive range of seas I purpose to regard as comprehending six provinces, since within them we can fairly reckon so many distinct centres of creation. The first and northernmost is the ARCTIC province, extending throughout that portion of the European seas included within the Arctic Circle. The second is the BOREAL province, including the seas which wash the shores of Norway, Iceland, the Faroe, and the Zetland Isles. The third is the CELTIC province, in which rank the British seas, the Baltic, and the shores of the continent from Bohuslan to the Bay of Biscay. The LUSITANIAN province includes the Atlantic coasts of the Peninsula. The MEDITERRANEAN province speaks its own explanation; the Black Sea is included in it. Lastly, the CASPIAN is a region now completely isolated from all the others.

Of these the four first named and the last are unquestionably distinct centres of creation; the Mediterranean and its dependencies are not so certainly entitled to that rank, and may possibly prove to be a chain of offsets from the Lusitanian area, just as the Baltic is of the Celtic, or the White Sea of the Arctic province. At the same time there is much to be said in favour of the more dignified view of the zoological importance of the great central sea; so much, that I will waive my prejudices against it, and treat it as an independent state.

The distribution of marine animals is primarily determined by the influences of climate or temperature, sea-composition and depth, in which pressure, and the diminution of light are doubtless important elements. All these may be combined so as to complicate the character of the fauna of a particular province. This appears to be especially the case in the Arctic seas, as I shall have hereafter to insist upon with much stress. The secondary influences modifying the action of the primary ones are many. Thus *the structure of the coast*, so far as the mineral character of its rocks is concerned may seriously affect the distribution of particular tribes. Whole families of marine animals depend for their subsistence on the presence of sea-weeds, and of the creatures that feed upon them. Yet all kinds of rocks are not favourable to the growth of weed, and tracts of sand may be wholly free from marine vegetation, or when giving support to sea-plants, cherish forms, adapted for the subsistence of peculiar animals only. Consequently whole tribes of beings may be present on, or absent from, a range of coast, according to its geological, or rather mineral structure, although every other condition be perfectly favourable to their propagation. And, what is more important, the course and diffusion of whole tribes may be restricted within areas far more limited than their capabilities for enduring elemental or bathymetrical conditions warrant, in consequence of the barrier interposed to their spreading through a sudden change in the structure of the land and sea-bed. The diffusion of burrowing marine invertebrata must be very seriously affected by such changes. Thus many shell-fish bore only in limestones, or rocks containing abundance of lime; a very ordinary difference in the nature of the coast must determine their presence or absence.

The *outline of a coast* has great influence in regulating the diffusion of species. A much indented region is very favourable to submarine life, a straight and exposed coast-line usually unfavourable, though there are a few creatures which delight in the dash of the waves, and hardily—though some of them are small and exceedingly delicate—brave the full force of the ocean-storms; reminding us of those sturdy people not uncommon in this stormy life, who thrive best in troubles, and feel happiest under conditions that make most men miserable.

The *nature of the sea-bottom* determines, to a great extent, the presence or absence of peculiar forms of shell-fish and other invertebrata, and of fish also, since, according to the food, so is the distribution of the devourers. We find very different creatures brought up by the line, net, or dredge, according as the bottom is of mud, sand, gravel, nullipore (coral-like sea-weed), broken shells, loose stones or rock, and the gradations of their intermixtures.

Tides are also modifying influences, and the extent to which they rise and fall is most important in determining the presence or absence of the species inhabiting the littoral zone. The shape and size of testacea found in tideways are very considerably influenced by their situation, and it is in such localities we seek with most success for the curious and beautiful sea-jellies (*medusæ*), whose fragile frames seem often to delight in sporting amidst the agitated waters.

Currents, besides their agency as modifiers of climate, act as means of transport, and, perhaps above all other causes, are influential in determining the diffusion of marine animals and plants, since, through their help, the germs and larval states of numerous creatures which eventually become fixed and stationary may be carried from district to district, and rapidly extended over vast areas. Even fixed creatures, when attached to bodies, such as masses of wood, capable of easy transport, may have their range materially enlarged by the same cause.

The influence of *climate* is conspicuously manifested in the diminution of the number of genera and species of marine animals in the European seas, as we proceed from south to north; this decrease we can scarcely attribute to other cause than the diminution of temperature. In the warm waters of the southern provinces, whether mediterranean or oceanic, the variety of types and the abundance of kinds ranged under them are equally multiplied; in the colder waters of the north, the forms are not so varied, nor are the species so numerous, though, as if in compensation, the number of individuals is so great as to prevent inconvenience from the comparative scarcity of kinds. In both vertebrate and invertebrate divisions of the animal kingdom this is manifest. We may exemplify the fact by reference to the best investigated sections of each. Thus, whilst the number of generic types of fishes in the Mediterranean region is 227, in the British seas we have only 130, and in Scandinavia there is a still further decrease to 120; of mollusca, in the first-named region there are 155 genera, in the second 129, and in the third 116. The number of species of fishes in the Mediterranean seas is 444; in the British seas 216; in the Scandinavian seas 170; and of marine mollusca (exclusive of Nudibranchiata and Tunicata, data for computing which tribes are insufficient), Mediterranean 600, British 400, and Scandinavian 300.

But climate alone is not the only cause of change in our course from north to south. The changes in the geological structure of the European shores are frequent and striking in that direction, and affect materially, or rather determine the physical aspect of the coasts, and the conformation of the neighbouring sea-bed. Geologists have to deal not merely with the land as it is exhibited above water; they must prosecute their science in the recesses of the sea, and trace in the depths and shallows of its floor the continuations of the plains and hills, and valleys of the contiguous lands, and seek for an explanation of its inequalities in the same gradual changes and sudden cataclysms to which their undulations, and levels, and ravines have owed their origin. These variations in the form of the surface, whether of dry, or of submerged land, importantly affect the distribution of living creatures, now furthering their progress beyond the regions to which they strictly appertain, now arresting their diffusion, and restricting them within limited areas far more circumscribed than the extent of climatal conditions, for which, were there fitting ground to favour their range, they are adapted by their organization.

The *composition of the waters* in which aquatic animals live, is a most important influence in its effect on their distribution. The degree of saltness or freshness determines the presence or absence of numerous forms of both fishes and invertebrate animals. Within the European area unusual conditions of this influence are manifest in the most northern, and a part of the most southern provinces. In the Arctic region, where unquestionably the small number of testacea in the shallows is in great part due to the comparative freshness of the upper layer of waters; in the Baltic Sea, where the waters are entirely modified; in the Black Sea, where the phenomena of the limited and peculiar fauna are in part determined by the peculiar character of this portion of the Mediterranean basin, modified as it is by its nearly complete isolation, and by the great rivers that flow into it; and in the Caspian, where the waters are of a nature very different from that of the ocean. In many confined localities, as in the lochs of Scotland, and the fiords of Norway, also in many estuaries, the surface waters may be fresh, or nearly so, whilst their depths are as salt as the open ocean, so that in the same place we may have creatures organized for very different states of sea-composition living not merely in the immediate neighbourhood of each other, but even, as it were, superimposed. I was once greatly struck with this fact, when dredging in the Killeries, along with Robert Ball and William Thompson, an arm of the sea in the wild and rocky district of Connemara, in Ireland. The depth was some fifteen or twenty fathoms, and the creatures inhabiting the sea-bottom were characteristically marine. When taken out of the water, they seemed to be unusually torpid, and it was in vain we placed them in vessels filled with the element of their native bay in order to tempt them to display their variously-shaped delicate organs. The cause of their languor soon became evident, when we remarked a fisherman dipping a cup into the water by the boat-side for the purpose of procuring some to drink. The uppermost stratum of the narrow and lake-like bay was purely fresh, or nearly so, derived doubtless from the numerous streamlets flowing into it, and from the rain, over-sufficiently abundant in that mountainous and picturesque district. The mollusca and radiata drawn from the salt waters beneath, became convulsed and paralysed in their involuntary ascent through the fresh waters above. They were more dead than alive when we placed them in basins, and none the livelier for having a new supply of water given them taken from the surface of the sea. Yet whilst these truly marine creatures were living and thriving below, numerous forms of entomostraca, incapable of enduring the briny fluids of the depths, might be sporting in the lighter and purer element above. This phenomenon, which I have often observed since, suggests the possibility of a mode of destruction of fishes which would aid in explaining the peculiar aspect not unfrequently presented by the fossil remains of those animals. In many places where petrifactions of fishes are found, their bodies are observed to be more or less contorted and convulsed. Many marine fishes when suddenly plunged into fresh water—and this is the case also with numbers of marine invertebrata—die rapidly, almost instantaneously, in convulsions, their bodies becoming suddenly stiffened, their fins spread and beautifully displayed. I have availed myself of this method of *piscicide* when desirous of obtaining sea-fishes in the state best adapted for delineation. Now it is not improbable that fishes of strictly marine habits, and incapable of enduring sudden immersion in fresh, or nearly fresh water, when too eagerly pursuing their prey, or too timidly flying from their pursuers in localities such as those I have referred to, might suddenly rise from the salt into the fresh water, be as suddenly paralyzed, and precipitated in their convulsed attitudes into the abyss below, where, sinking rapidly in the soft mud of the sea-bed, their remains would become enveloped and potted, before the numerous animals that creep and swim, watchful for carrion everywhere in the habitable depths of the ocean, had become aware of the neighbourhood of such acceptable prey.

The influence of *depth* is everywhere manifest in the European seas, for everywhere we find creatures whether animal or vegetable, distributed in successive belts, or regions, from the margin of the high water-mark, down to the deepest abysses, from which living beings have been extracted. Peculiar types inhabit each of the zones in depth, and are confined to their destined regions, whilst others are common to two or more zones, and not a few appear to have the hardiness to brave all bathymetrical conditions. Nevertheless, so marked is the *facies* of the inhabitants of any given region of depth, that the sight of a sufficient assemblage of them from some one locality, can enable the naturalist to speak at once to the soundings within certain limits, and without the aid of line or plummet. Throughout the oceanic portion of the seas of Europe, four distinct and well-marked zones of life succeed each other. The first of these is the *littoral* zone, equivalent to the tract between tide-marks, but quite as manifest in those portions of the coast-line where the tides have a fall of a foot or two, or even less, as in districts where the fall is very great. This important belt, which is inhabited by animals and plants capable of enduring periodical exposure to the air, to the glare of light, the heat of the sun, the pelting of rain, and often to being more or less flooded with fresh water, when the tide has receded, claims many genera as well as species peculiar to itself. These again are not distributed at random within the littoral space, but are ranged in sub-regions, which may be traced on rocky shores when the tide is out, even by the most inexperienced eyes, forming variously-coloured belts, banding the base of the land. Their peculiarities will be best pointed out when we treat of the features of the littoral zones as exhibited in the several European provinces. Succeeding this great shore-band we have the region of sea-weeds—the *laminarian zone.* It extends from the edge of low-water to a depth varying in different localities, but seldom exceeding fifteen fathoms. The *laminarian zone* is itself divided into sub-regions, marked by belts of differently-tinted algæ. It claims a numerous population of animals peculiar to itself, and is the chosen residence of fishes, mollusks, crustaceans, and invertebrata of all classes, remarkable for the brightness and the variegation of the patterns of their colouring. This region, above all others, swarms with life, and when we look down through the clear waters into the waving forests of broad-leaved tangles, we see animals of every possible tint sporting among their foliage, darting from frond to frond, prowling among their gnarled roots, or crawling with slimy trail along their polished bronzy expansions.

To the *laminarian* succeeds the *coralline zone*, wherein the horny plant-like polypidoms of hydroid zoophytes delight to rear their graceful feathery branches, whose flowers are animals rivalling plants in symmetry and beauty. This region has a wide extension, well on to some thirty fathoms or more in most places, commencing at the termination of the zone of sea-weeds, especially at that portion of the latter where the coral-like nullipores, vegetables simulating minerals in figure and consistence, abound and furnish a ground well fitted for the spawning of fishes. Here we have great assemblages of marine animals, both vertebrate and invertebrate, but plants are "few and far between." Last and lowest of our regions of submarine existence is that of *deep-sea corals*, so named on account of the great stony zoophytes characteristic of it in the oceanic seas of Europe. In its depths the number of peculiar creatures is few, yet sufficient to give a marked character to it; whilst the other portions of its population are derived from the higher zones, and must be regarded as colonists. As we descend deeper and deeper in this region its inhabitants become more and more modified, and fewer and fewer, indicating our approach towards an abyss where life is either extinguished, or exhibits but a few sparks to mark its lingering presence. Its confines are yet undetermined, and it is in the exploration of this vast deep-sea region that the finest field for submarine discovery yet remains. Such is the general subdivision of the sea-bed as exhibited in the European seas; in the Mediterranean, however, as might be expected, when we consider the peculiar conditions under which that great land-locked basin is placed, there are peculiarities in the distribution of both animal and vegetable life which require special consideration, and which we shall examine when we come to the description of the Mediterranean province.

CHAPTER II.

ARCTIC PROVINCE.

OF the six centres of creation shared by the European seas, one of the least prolific in number and variety of species, is the Arctic province, within which are included the snowy and inhospitable islands of Spitzbergen and Nova Zembla, the northern coasts of Russia, the coasts of Finmark, and, though rather as an intermediate and bounding state, the greater part of the shores and islands of Nordland—in short, all those continental portions and insular dependencies of Europe that lie within the Arctic Circle. In the strictest sense, the extensive, though barren, islands of the Arctic Ocean, and the very northernmost points of the continent only can claim to exhibit the true and typical features of the Arctic province. We should probably be justified in comprehending within it the northernmost shores of Iceland; for the margin of this region, which is itself much more extensive than its European portions indicate, extending, as it does, through the icy seas of Arctic America, becomes more and more southern towards the western side of the Atlantic, and in the New World impinges on the shores of Labrador and Newfoundland. From its easternmost bound in Europe it is extended, moreover, along the whole of the northern coastline of Asia, onwards to the Icy Cape. It would appear also to be intimately connected as a zoological province with the seas of Kamtschatka and Ockhotsz, which, as we shall see hereafter, share in its fauna, carried down on the western side of the North Pacific to equal latitudes, and with a similar distribution to that which it exhibits on the western side of the North Atlantic. Unlike all other marine zoological provinces, unless there be an exception in the Antarctic regions, it is continuous, and belts the globe around the North Pole; not, however, with an even circular boundary, but as we have seen, with a variable and undulating edge. Wherever arms of the sea branch from it, they carry its fauna and flora with them, to the exclusion of all other populations. Of this the White Sea presents a most instructive example, for extending inland from that portion of the Arctic province where the Arctic Circle is rather without than within its bounds, this great offset of the Arctic Ocean carries its peculiar fauna and flora into the heart of the land, and beyond their natural bounds, placing them, as it were, side by side with the population of the Gulf of Bothnia, which, extending from the Baltic, itself an arm of the Celtic province, carries the remnants of a fauna exclusively Celtic, to the verge of the Arctic Circle.

The region in which the sea is permanently frozen

in winter is equivalent to the typical, or main portion of this Arctic province, and the line of the winter ice to the boundary of that section of it. The shores of Norway, or rather Finmark and Nordland, from the North Cape to the Arctic Circle, are debatable ground, sharing in the features of the Boreal province, yet presenting besides such marked indications of the Arctic fauna, that I think it best to include them here.

In this realm, where King Frost holds despotic sway, where the long year is longest, and yet consists of but a day and a night, where man shrivels into a dwarf, and has plainly no just claim to a dominion, the power of the Creator has called forth the mightiest and the minutest of the inhabitants of the ocean. The kinds of living things there are few, but those few display peculiarities which mark them as members of an assemblage of organized shapes, whose home and birth-place are in the icy seas. The polar regions are not negations so far as animated nature is concerned. Bleak and desolate, cold and dreary as they are, they do not constitute merely the boundaries of the regions of life; they have their own peculiar beast-people on land and in the sea. They are no deserts whence the Caller-forth of life has been absent; among the glassy icebergs there has been a genesis.

> " That sea-beast
> Leviathan, which God of all his works
> Created hugest that swim the ocean stream,"

first rose amongst the floating mountains of crystal ; and many a beauteous bubble endowed with life first sported in the chilly waters washing their polished sides.

The physical influences that affect the distribution of marine animals in the Arctic province are various, and somewhat complicated. The sea bounding the northern extremity of Europe is, in the main, deep, and very deep in parts. Between Spitzbergen and West Greenland twelve hundred fathoms of line have not reached the bottom. There are fathomable depths enough, however, and sufficient ground within the range of the laminarian, coralline, and deep-sea coral zones to cherish the development of animal and vegetable life, if unfavourable influences did not interfere. Cold is the great arrester of organization in these northern regions. Its influence is chiefly exerted in the littoral and laminarian zones, partly through the low temperature of the air, partly through the cold waters of the Arctic current flowing from the eternal ice of the pole onwards towards the south-west. But the unfavourable effects of the surface waters are modified by the higher temperature of the depths, for in the Arctic seas (and also in the Antarctic) the temperature increases as we descend, contrary to what takes place in the seas of temperate and warm regions. The consequence is, that instead of the marginal zones of these seas being most favourable to the development of animal and vegetable life, they are the most unprolific, and we have to descend into the depths to find an abundance of ground-living creatures, which moreover appear to range much deeper in high latitudes than they do in more favourable climes. Their bathymetrical range will, in the end, probably be found to accord with the breadth of the stratum of water of the temperature they require. It is the warmer currents flowing from the south northwards, and passing beneath the cold waters of the arctic current, that originate this distribution of temperature and animals in depth. A very important fact is this—for, as we know from observation, the animals of the depths are members of the fauna of the Boreal or next southern province; whilst it is in the shallows, or along the littoral and laminarian belts on the coast, or in the colder upper waters, unfavourable as they are to life, that we find the characteristic and peculiar members of the Arctic province. The presence of the former is, in all probability, due to their diffusion northwards by the under current. The paucity of numbers of the marine creatures inhabiting the higher zones may also be in part dependent on the composition of the waters of the Arctic seas, for their upper stratum is less salt than in seas more to the south; whilst the greater saltness of the under layer, taken in connection with the exceeding clearness of the waters, through which the bottom and the shells upon it are plainly visible, even at a depth of eighty fathoms, go far to favour the development and extraordinary bathymetrical range of the animals inhabiting the depths. For light is an influence of great power in the development of marine life. On the other hand, we may attribute to the general deficiency of light during a great part of the year, the dulness of hue which is so marked a feature of Arctic animals.

The existence in Spitzbergen of great beds of clay, forming cliffs one hundred feet high, and an extension of land of considerable dimensions,[*] containing fossil shells, all of Arctic species, and indeed, corresponding exactly with the characteristic inhabitants now living in the Arctic seas, shows that the existing zoological condition of the Arctic province has been of long standing, and dates back, in all probability, from the Pleistocene epoch—that which immediately preceded the present. Yet, at an infinitely distant period in time, very different conditions of climate must have prevailed in this now barren and inhospitable region; for beds of coal, and traces of an extensively developed vegetation, and limestones abounding in organic remains, form parts of the structure of that ice-bound island,[†] which is mainly made up of palæozoic rocks of sedimentary origin. The strange tower-like mountains and spindle-shaped peaks, that call forth expressions of admiration and wonder from all who have sailed along its rocky shores, mostly owe their ec-

[*] Keilhau.   [†] Robert.

centric outline to the crumbling away and weathering of great beds of conglomerate and brecchia.

Whether it is that the paucity of objects for observation induces attention to such as are seen, or that the cold air sharpens men's wits, voyagers to the Arctic seas, not being professed naturalists, have paid much more attention to the animals which inhabit them, and described those they have met with much more intelligibly than travellers in warmer and more favoured climates. Most of the writers on northern latitudes give some account of their natural history. In the region under review, the early adventurers in the whale fisheries did not omit to observe, with considerable care, the characters, differences, and habits of the animals they pursued, and, at the same time, were not so blinded by the magnitude of their prey as to pass without notice some of the more striking among the minute organisms vivifying the polar waters. Among other places, Spitzbergen, the delineation of the fauna of which is of great consequence in the history of the European Arctic region, since it is clearly the part of it where we should expect to meet with the type of that fauna, has fortunately not been neglected. In the expedition towards the North Pole, undertaken in 1773, under the charge of Captain Phipps, important data for the determination of the natural history of Spitzbergen were collected, not at hazard, but with evident judgment, and a clear understanding of their value. These were published in a strictly scientific shape. In the most valuable and interesting "Account of the Arctic Regions," by Dr. Scoresby, Spitzbergen, well known, through personal research, to that now eminent author and philosopher, receives full attention, and its natural history features are carefully noted. Although a practised naturalist would scarcely fail, if he visited it, greatly to enlarge the published catalogues of the inhabitants of this barren but picturesque island, the enumeration given by the author cited, is of such a character, that we cannot doubt that it includes the main features of its marine fauna.

Were it not for the peculiarities of its zoology, Spitzbergen might rear its spiry peaks for ages unscanned by human eyes, and no voice of living man be heard among its frozen solitudes. But strange and bulky creatures, whose organization and habits constitute their links between the land and sea, throng in these dreary regions, and have chosen them for their own. Seals of various kinds are gathered there in herds; the fearless and bulky walrus crowds on the icy edges of the desert island, and with its human head and powerful tusks, seems as if it were the guardian spirit of the enchanted wastes. Than this animal there is none more characteristic of the Arctic province. Not less so, however, are the mighty whales, who career through the waters of the Arctic Ocean :—

> " Enormous o'er the flood, Leviathan
> Looked forth, and from his roaring nostrils sent
> Two fountains to the sky, then plunged amain
> In headlong pastime through the closing gulf."

Headlong pastime truly! for down into the ocean's abysses, four hundred fathoms and more can the gigantic monster plunge when seeking for his food, or flying from his pursuers.

The Greenland whale (*Balæna mysticetus*) is, or rather was, the grand feature of these seas, and the great temptation to adventurous voyagers. This mighty beast of ocean, whose bulky body reaches a length of sixty feet, has been the source of much wealth, and the theme of many fables. All his bigness was not sufficient to content the lovers of the marvellous, and his dusky skin had needs have been made of India rubber, to have borne the stretching endured by it in the writings of some of its wondering describers. But the whale has his diminishers as well as his magnifiers, and all his bigness cannot save him from destruction. In the year 1814 the whale-fishers killed no fewer than fourteen hundred and thirty-seven whales, and one lucky skipper had the marvellous fortune to bag forty-four whales for his own share in the neighbourhood of Spitzbergen. But whale's blubber cannot put up with incessant persecution any more than human flesh, and the golden mine of the Spitzbergen seas has been exhausted. Well on to two centuries and a half have passed away since the whale fishery was commenced by English enterprise in the Spitzbergen seas, and rich, indeed, must have been the products of the venturers, since, so late as the year I have just noticed, the fish were abundant. When Mr. Scoresby, in the year 1820, published his "History of the Whale Fishery," the whales still frequented those seas in considerable numbers. In 1840 they had left them apparently for good and all. Professor Jameson, in a note on the state of the fisheries, published in his valuable journal during that year, states that "the whale fisheries between Spitzbergen and Greenland are abandoned. Fishers now prefer Davis Straits, Baffin's Bay, or the seas to the east of Greenland." Davis Straits is now likely to be deserted, and the whale fishery is diminishing, the number of whale-ships decreasing yearly. Thus has the activity of man done much towards rendering one of the mightiest of living animals well nigh extinct. If this fishery be pursued for a century longer, the Greenland whale may take its station with creatures that have been.

The rorquals, or, fin-whales, still hold their places in these seas. Their rapid movements defy the efforts of human enemies, though probably all their activity would be of little avail were they sufficiently remunerative for the trouble of killing them. The mightiest of all leviathans, the *Balænoptera boops*, is among their number, growing to the vast length of one hundred feet and more. The *B. boops*, *musculus, physalis,* and *rostrata* are all inhabitants of the Spitzbergen seas.

This, too, is the realm of the sea-unicorn, a creature quite as strange, but not as fabulous as the terrestrial animal, whose golden image is so familiar to us in England. The narwhal (*Monodon monoceros*) derives his popular misnomer from the enormous tusk, projecting from its upper jaw, the fellow tooth being undeveloped. What purpose this formidable weapon serves, has not yet been clearly made out, and the balance of opinion inclines to decide that it is no instrument of defence, but rather a mark of superior dignity; a sceptre wielded by the male sex alone, to assert in the most prominent fashion their superiority over their gentler mates. Be that as it may, this curious creature is certainly one of the most interesting and peculiar inhabitants of the Arctic seas.

Of the dolphins that frequent the Spitzbergen seas, the "White Whale," or Bjeluga (*Delphinus leucos*) is the one chiefly turned to account in the North. But they are much more numerous near the continental portions of the Arctic province than in its remoter abysses.

A striking feature is the paucity of fish at Spitzbergen. In Scoresby's zoological summary but six species are indicated, to four of which names are attached. Phipps met with two kinds of fish only, although he seems to have made diligent search for them, and to have used the trawl freely with that view. These were the green cod or *sei* (*Pollachius virens*), abundant enough in Nordland and Finmark, and a little sucker (*Liparis vulgaris*). Scoresby notices and describes a more characteristic fish, in the Greenland shark (*Læmargus borealis*), a large animal, twelve or fourteen feet in length or more, and six or eight feet in circumference. It is harmless to man, but an enemy of whales, biting and tearing its superior monsters when alive, and eating them up when they die, gorging itself with blubber like an Esquimaux, or other northern person, scooping hemispherical pieces, each as large as a man's head, out of the whale's body, and swallowing as much as ever it can, until it has so filled itself with its dinner that it has no place wherein to stow away any more; heeding no annoyance, not even the stab of knives at dinner-time, and contenting itself with a fasting diet of small fishes and crabs on those days when whale's beef is forbidden, because not to be procured.

This scarcity of species of fish does not hold good throughout the Arctic province, for in its southern and transitional portions there are not only large fisheries established, but numerous kinds taken. Travellers to the North Cape have often remarked the abundance of fish seen in the clear waters of the Finmark seas. Thus Capel Brooke[*] observes that the different levels of the sea in the bay of Hammerfest seemed swarming with different kinds; the

[*] Tour in Lapland, &c.

upper layer of the water being thronged with young cod, the middle depths abounding with sei, whilst on the floor of the sea, studded with sea-urchins and star-fish, and but rarely with shell-fish, huge plaice and halibuts might be seen gliding or lurking. The sei fishing is indeed a chief branch of Finmark trade. The capelan is stated by Nilson to visit the shores of Nordland and Finmark in spring, for the purpose of depositing its eggs.

Crustacea seem to be more abundant at Spitzbergen, and more than half-a-dozen kinds are mentioned by Phipps and Scoresby, which number, when we bear in mind the proportion of species scarcely taken note of to those that attract attention, must indicate a very considerable list, and of late years, not a few remarkable new forms have been described from this quarter. Ten kinds of mollusca are enumerated. Of these, two are bivalves (*Mya truncata* and *Hiatella rugosa*), both of common forms, and ranging in abundance to the British seas; three gasteropodous univalves (a *Chiton*, a *Buccinum*, and a *Margarita*); two pteropods; one cuttlefish, taken abundantly from the stomach of narwhals, and apparently constituting their favourite food; and three Ascidians (enumerated as *Ascidia gelatinosa* and *rustica*, and *Synoicum turgens*); the circumstance of these last curious but unattractive animals receiving attention, shows how zealously our voyagers laboured. In an excursion on shore Mr. Scoresby appears to have searched for mollusca, but was unsuccessful and found none. As he notices, however, the presence of *Fucus vesiculosus, Laminaria saccharina, Alaria esculenta* and other large sea-weeds of the Laminarian zone, I do not doubt that a minute search, rightly directed, would bring to light some of the species of *Rissoa* and *Lacuna* which inhabit those plants on the coasts of Greenland. Of late years the *Trophon scalariforme*, an elegant kind of whelk, found abundantly fossil in our British drift, and common on the coasts of Labrador, has been taken in the Spitzbergen seas, but does not range as far south as Norway on this side of the Atlantic.

Small as they are, two little pteropods (*Limacina arctica* and *Clio borealis*) are among the most important inhabitants of these seas, since they constitute no inconsiderable part of the food of the whale. Like other members of their family they are swimmers, active and graceful in their motions, moving through the water by means of their wing-like muscular fins, and seeming, as has often been remarked, the butterflies of the sea. The *Limacina* is covered with a spiral shell of extreme tenuity, and elegant curvature; the *Clio* has no such appendage. Mr. Scoresby remarks of the former, that it is found in immense quantities near the coast of Spitzbergen, but does not occur out of sight of land; and of the latter, that though met with in vast numbers in some situations near that island, it is not distributed generally throughout the Arctic seas.

Yet more noticeable are the jelly-fish, or medusæ, of these regions, which, indeed, along with the cetacean giants, who either directly or indirectly derive their subsistence from them, constitute the main and characteristic zoological feature of the Arctic province. The minuter medusæ throng the icy seas in countless myriads, and their abundance and exceeding beauty have attracted the attention of all northern voyagers. Great shoals of them are met with discolouring the water for a vast extent. Scoresby observed that the colour of the Greenland sea varies from ultramarine blue to olive green, and from the purest transparency to striking opacity, appearances which are not transitory but permanent. The green semi-opaque water mainly owes its singular aspect to minute medusæ, and infusorial animals. It is calculated to form one-fourth part of the surface of the Greenland sea, between the parallels of 74° and 80°. It is liable to alterations in its position from the action of the current; but is always renewed, near certain situations, from year to year. Long bands or streams of it, having a direction of north and south, or north-east and south-west, sometimes extending two or three degrees of latitude in length, and having a breadth of from a few miles to fifteen leagues, are met with. The whales throng in this muddy water, for to them it is good wholesome soup, nourishing enough, as may be judged from the curious calculation of the observant voyager I am quoting. "The number of

Medusæ," writes Mr. Scoresby, " in the olive-green sea was found to be immense. They were about one-fourth of an inch asunder. In this proportion, a cubic inch of water must contain 64; a cubic foot 110,592; a cubic fathom 23,887,872; and a cubical mile about 23,888,000,000,000,000! From soundings made in the situation where these animals were found, it is probable the sea is upwards of a mile in depth; but whether these substances occupy the whole depth is uncertain. Provided, however, the depth to which they extend be but two hundred and fifty fathoms, the above immense number of one species may occur in a space of two miles square. It may give a better conception of the amount of medusæ in this extent if we calculate the length of time that would be requisite, with a certain number of persons, for counting this number. Allowing that one person could count a million in seven days, which is barely possible, it would have required that eighty thousand persons should have started at the creation of the world" (the writer refers to popular, not geological reckoning), "to complete the enumeration at the present time." The microscopic thread-like infusorials, called *Galionellæ*, appear to have a considerable share, as well as the minute jelly-fishes, in producing the discoloration of the water. Judging from the imperfect figures given in the plates to the "Account of the Arctic Regions," the little animal to which the name *Appendicularia* has been applied, probably plays no small part in producing this phenomenon, especially where the tint of the water may be inclined to red. The minute globular animalcule, with a dark-coloured tail, advancing by a curious zigzag motion, figured in Plate 16, No. 19, of the work referred to, seems to me to be this anomalous creature. I once examined a track of reddish water off the Zetlands, and found the *Appendicularia* to be the cause of the colour. Its true position in the animal kingdom has just been made out, and Mr. Huxley has established its claims to a higher rank than that held by the jelly-fishes with which it keeps company.

The members of the medusa tribe, which appear to abound most in the Arctic seas, are ciliograda, creatures which are, for the most part, more or less spherical in shape, or else simulate strips of ribband, always as transparent as the purest crystal, and moving through the water by means of variously arranged bands of thread-like hyaline fins, which, as they flap, all in each long row keeping exact time, decompose the rays of light, and glitter with the hues of the rainbow. More exquisitely beautiful creatures than these *Beroidæ* (for so the tribe is called) do not exist among all the wondrous beings that people the sea. The elegance of their shapes is equalled by the grace of their movements; and when the prismatic lustre of their bands of cilia marks the course of their crystal bodies, as they swim with gentle motion through the water, they seem as if they were diamonds endowed with life. Some, such as the *Beroe cucumis*, one of the most characteristic of the northern forms, yet having a wide range to the south, although in fewer numbers, are tinged with a charming amethystine blush. This is the "Fountain-fish" of the early voyagers to Spitzbergen, who, mistaking the cause of the eight bands of iridescence, gleaming along the sides of its body, fancied they were so many rivulets of lustrous water. Another, the *Cydippe* (of which the species represented by Scoresby, is probably my *Cydippe Flemingii*, and not the true *pileus*), is furnished with two long pinnated filaments lodged in sigmoid cavities, one on each side of its stomach. By these filaments it can moor itself, as well as guide its path through the waters, retracting and expanding them at pleasure, and, when in rapid motion, usually withdrawing one or the other alternately. A third kind, the *Mnemia* (of which Scoresby's figures, pl. xvi. fig. 3 and 5, are most probably imperfect representations), has no filaments, and in general contour resembles the *Beroe*, but differs in having its sides developed anteriorly into great flaps, or swimmers, and the possession of four lanceolate tentacula surrounding its mouth, which, like the *Beroe*, it carries downwards when swimming, whilst the contrary position is customary with *Cydippe*. These delicate and beautiful Ciliograda are abundant throughout the Arctic seas, and seem to have attracted the attention of all the voyagers to Greenland. They range to our own shores, and south of them, but are comparatively scanty, and are scarcely observed even by our fishermen, who, when they notice them, designate them as "spawn."

It appears probable that two of the most anomalous of the swimming animals of the Ocean, species of *Sagitta* and *Briarea*, are abundant in the Arctic seas, though unrecorded by name. The former is, as its name implies, an arrow-shaped creature; it is of exceeding simplicity of structure, and in shape resembles, as it were, a miniature draught of a Cetacean, for the regularly formed fin which terminates its tail, is transverse, and the general outline of its body bears out the comparison. It is very minute, and perfectly transparent, resembling an arrow of glass, and shooting through the water with the rapidity of a dart. Its only hard parts are the comb-like jaws with which its mouth is armed. The creatures figured by Scoresby in his Plate XVI. figs. 1 and 2, are evidently *Sagittæ*, a fact which does not appear to have been noticed by commentators. The *Briarea* is also a small and transparent creature of glassy texture; it is furnished with many lobes, each bearing two fin-like expansions at their extremities, and has two long tentacles which, being strengthened by cartilaginous rods, it can bend stiffly back on its body in a most dexterous fashion; its internal organization is of the simplest order. *Briareæ* were met with plentifully, and for the first time recognised in the Arctic seas, during the passage of the lost expedition of Sir John Franklin, through Davis Straits. I mention these two curious animals, the true systematic position of which is as yet doubtful and disputed, with some stress, since they have been met with in almost all parts of the ocean, central and southern, but have as yet not been put on record from these regions.

Zoophytes appear to be few and scarce, and those enumerated are not peculiar. Echinoderms, on the other hand, are not only plentiful, but so far as star-fishes are concerned, there is a marked and peculiar assemblage of species and even genera. The *Pteraster militaris, Ctenodiscus polaris, Ophiolepis Sundevalli, Ophiocoma arctica, Ophiocantha spinulosa,* and *Ophioscolex glacialis,* are star-fishes not known out of the Arctic province. *Comatulæ,* the curious and beautiful feather-stars, are, I am informed by Professor Goodsir, on the authority of a collector employed by him to dredge at Spitzbergen, so abundant in moderately deep water there, that their bodies frequently filled the dredge, to the exclusion of all other creatures. This is a fact of no small significance, when we recollect that the abundance of the remains of Crinoids in some ancient strata, has generally been regarded by geologists as supporting the notion of the prevalence of a warm climate within the British area at the time of their deposition.

The marine calcareous vegetable, *Nullipora poly-*

morpha, appears to be common in the Arctic seas; nor are the olive-coloured *Algæ* deficient.

The natural history of the coasts of Nova Zembla we know, from the researches of Von Baer, to combine the features of Spitzbergen with those of the continental shores of the Arctic Ocean; the former island presenting the characters of an arm of the mainland, and consequently possessing a greater number of both terrestrial and marine inhabitants. It is frequented by seal-hunters, who take here several valuable species (*Phoca leporina, barbata, groenlandica,* and *hispida*), as well as the walrus. The Bjeluga (*Delphinus leucos*) is also an object of search. The Greenland whale never strays as far as Nova Zembla, from which fact Von Baer infers, that the fishery carried on by Northmen in the ninth century, between it and the North Cape, must have been for the fin-fish (*Balænoptera*), comparatively difficult as that monster is to capture. It seems more probable, however, that the great whale retired from this sea, as it has lately retired from the Spitzbergen seas. In all other respects so far as marine mammalia are concerned, Nova Zembla resembles Spitzbergen. But nine species of fish, whether marine or anadromous, were met with, and of these two only served to play an important part in the fauna, *Gadus sarda* and *Cyclopterus liparis*. Small crustaceans, more especially *Gammari*, are abundant, and the vast number of aquatic birds, especially guillemots and gulls, are evidence that

THE EUROPEAN SEAS. 49

there must be abundance of living food for them in the surrounding waters.

The mollusks of Nova Zembla, and the neighbouring coasts of Russian Lapland, have been made known by Middendorff. The total number observed by that eminent naturalist and Von Baer, was sixty-eight species, of which forty were univalve testacea, and twenty-five bivalves, the remaining three being naked pteropods, or nudibranchs. In this number we have the minimum of species presented in a class by any truly marine province, and the entire assemblage represents molluscan life under the influence of the severest conditions. Fifty of these denizens of the Icy seas are identical with species found fossil in the drift beds of Great Britain or the south of Sweden. Between fifty and sixty range to the east coast of Arctic America, and considerably more than a third of the entire number reach Behring's Strait, or even range into the sea of Okhotsk. Among the forms which appear to range completely through the Polar seas, are the *Natica helicoides, Natica clausa, Cancellaria viridula, Purpura lapillus, Trophon scalariformis, Fusus islandicus, Terebratula psittacea, Pecten islandicus, Modiola modiolus, Mytilus edulis, Astarte elliptica* and *corrugata, Saxicava rugosa, Mya truncata, Mya arenaria,* and *Panopœa norvegica.* How strikingly does this assemblage remind us of the fossil fauna of the glacial epoch! Species that have, countless ages ago, deserted our British waters

E

are still found flourishing in the frigid recesses of the Arctic Ocean.

The uniformity of the fauna of the Arctic regions, is such, that in default of information from the European side of the ocean, we may turn, without much danger of error, to the seas of Arctic America. Systematic observations on the distribution of animal life in depth, are greatly to be desiderated from these seas; and, if ever (there are those who still hold hope) Sir John Franklin and his brave companions return to us, we may expect such information. My very dear friend, Mr. Harry Goodsir, sailed in H.M.S. Erebus as assistant-surgeon and naturalist. No more able or better qualified person could have been chosen for the scientific duties to which his attention was directed. He had already, though very young, gained a high reputation for his researches among marine animals, and had especially investigated the more critical and unpreservable tribes. He entered upon the dreary and dangerous voyage filled with scientific zeal and determined, among other inquiries, to prosecute a series of dredging observations, and to keep full records of the results. In a letter which I received from him when the ships were at Disco, on the west coast of Greenland (70° N. lat.), he dilates enthusiastically on the prospects of his Arctic studies, the promise held out by some observations he had already succeeded in making, and the zeal and delight with which all his companions, officers, and crew, entered into his pursuits. "Ever since I have begun work," he writes, "the officers have been exceedingly zealous in procuring animals for me, so that my time is completely occupied, almost day and night, for, from the constant light, and having generally lots of animals on hand, I am anxious that none should be lost. All are anxious to assist, down to the men, who have got several very good things for me. The boatswain is sometimes seen running after a specimen with the large net in hand." On the 25th of June (1845), when in Davis Straits, soundings were taken in forty fathoms, when a small dredge was put over. It brought up starfishes, echini, mollusca, crustacea, and annellida. Among the shells was a small *Terebratula*. On the 28th, they sounded in three hundred fathoms, and sank the dredge at that great depth. The bottom proved to be of greenish mud, and they had "a capital haul,—mollusca, crustacea, asteridæ, spatangi, corallines, a nondescript *Fusus*, Isopoda, and what is interesting to me, my genus *Alauna*, and your *Brissus lyrifer* (a curious sea-urchin), and some fine corals." The floor of the sea was composed of very fine green mud, which when placed under the microscope, appeared to be composed "of granitic particles." The next day they sounded in two hundred and forty fathoms, and met with the same green mud, but when, this time, it was placed under the microscope, it appeared to be composed of sandstone particles, with small fragments of shells, and of spines of *Echinus*, and *Spatangus*, mingled with great quantities of mucus.

In the "Fauna Groenlandica," of Otho Fabricius, there are forty-one species of marine fish enumerated, and more have been added of late years, by Kroyer, and other northern naturalists. The additions have been chiefly of purely Arctic forms. Several of those which Fabricius regarded as identical with more southern species, have proved to be distinct. In his list he indicates differences in the distribution of the species, a considerable number being confined to the southernmost parts of Greenland. These are exactly such as, on the European side of the Atlantic, fall into the southern or boundary portion of the fauna of its Arctic province. One fish preeminently plays a typical part in the Greenland Fauna; this is the capelan,— it is also, though not so abundantly, present in the Arctic fauna of Europe, and reaches the northern shores of Iceland. Of it I shall have to speak more fully in a future chapter. The lump-fish and the wolf-fish also have a prominent place in the Greenland fauna. The latter is in the habit of crunching into fragments strong shell-fish and crustacea by means of its powerful jaws. A curious and ingenious attempt has lately been made to refer[*] the fragmentary condition of the shells contained in the clays of the pleistocene formation, to the voracity

---

[*] Mr. Craig in "Geological Journal" for 1850.

and destructive power of this formidable animal, which, whatever may be thought of the speculation, in all probability frequented our seas in vast numbers, during the glacial epoch, and had doubtless the same predacious habits then as now.

The number of Annellides procured in the Greenland seas is large, and the researches of Oersted would lead us to estimate the development of the dorsibranchiate tribes to be greater in the Arctic than in more southern seas. But in the present state of science, the sea-worms have not been sufficiently investigated in any part of Europe to afford a just basis for comparison. Much less have they been examined out of Europe.

In the enumeration of Scandinavian mollusca, by Professor Lövén, a certain number of testaceous species are mentioned as not found south of Finmark and Nordland. These may be regarded as characteristic animals of the Arctic province, as it is presented in its southern and continental portion. The Pteropods *Clio* and *Limacina*, already mentioned, are among them. Of Gasteropoda there are *Philine scutulum, Trophon harpularium, Trophon Gunneri, Cancellaria viridula, Lamellaria prodita, Natica clausa* and *aperta, Lacuna labiosa* and *frigida, Margarita cinerea, Scissurella angulata, Acmœa rubella,* and *Chiton nagelfar*, the largest European chiton. Of Brachiopods, there is *Terebratula septigera*. Of Lamellibranchiate bivalves, there are *Pecten imbrifer* and *Groenlandicum, Modiolaria lævigata* and *Mactra ponderosa*. These are either species described for the first time, or old ones common to both sides of the Atlantic. Thus *Trophon harpularium, Cancellaria viridula, Natica clausa* (and probably also *N. aperta*), *Margarita cinerea, Pecten groenlandicum, Modiolaria lævigata,* and *Mactra ponderosa,* not only range to the strictly Arctic shores of America, but most of them descend as far on the western side of the N. Atlantic, as the banks of Newfoundland, and the neighbourhood of Cape Cod. This is the case, also, with *Scalaria groenlandica,* and with *Terebratula septigera* and *Astarte corrugata,* two bivalves recorded from Finmark only, in the northern fauna, but known under very exceptionable circumstances farther to the south. It is a striking and important fact, that several of these species—so widely diffused on the American coast, whilst, on the European they are restricted to the Arctic circle—ranged, at the epoch of the drift, as far south as the middle of England, and the south of Ireland; their fossil remains, undistinguishable from recent specimens, are found in the strata of the drift epoch in numerous British localities at the present day. This fact cannot be too strongly impressed on geologists, many of whom have an impression that there is no marked difference between the fauna of the drift, and that of the British seas at present, because the species of shells found in the former are species still living. But the presence of three or four such species—to find which alive in the European seas, we must now travel to the bounds of the Arctic Ocean—combined with the absence of the great body of Celtic species, has a significance of deep import. Not even on the verge of the Arctic province are we to seek for the analogue of the fauna of the drift, but within its strictest bounds. Of this, however, more hereafter.

The number of mollusca recorded from the coasts of Finmark, affords an indication of the degree of fertility of that region in species. There are three Cephalopods, of which one is peculiar and new; three Pteropods, one of them ranging as far south as Scotland; four Nudibranchs, two of them peculiar; sixty-six univalve Testacea, of which thirty-six range as far south as the British seas, or farther; four Brachiopoda, one of which is new; and forty-five ordinary Bivalves, of which all but eight range to the British seas; making a total of one hundred and sixty-nine species. This number is considerably larger than that of the Greenland molluscan fauna, which amounts to one hundred and thirty-four species. The difference is due to an infusion of species advancing from the south, along the continuous shores of Norway, in the one case, whilst the Arctic fauna is isolated on the other. The Greenland number is therefore a truer expression of the Arctic molluscan fauna (exclusive of Tunicata, which I have not counted in either case), than the Finmark number. The authorities I follow, with some slight revision, are H. P. Müller for Greenland, and Lövén for Finmark.

About fifty univalves and bivalves are enumerated among Greenland testacea, which do not appear in the European lists; but this number, since most of them are said to be new, and many are known only by very brief descriptions, will probably, on close investigation, require considerable reduction. On the other hand there are about fifty-four testacea common to Greenland and the Scandinavian seas, and out of this number, thirty range to the Scottish seas. It is a very remarkable fact that the species of shell-fish common to Greenland and Finmark are not all inhabitants of deep or moderately deep water, but that among them we find periwinkles (*Littorina rudis, var., Groenlandica*, and *Littorina retusa*), the dog-whelk (*Purpura lapillus*), and the little *Skenea planorbis*, all of which are inhabitants of the belt between tide-marks; also the tortoise-shell limpet (*Acmæa testudinalis*), the common mussel (*Mytilus edulis*), and species of *Margarita* and *Lacuna*, whose dwelling is at the margin of low water, or in the belt of weed immediately succeeding. That these littoral mollusks indicate by their presence on both sides of the Atlantic, some ancient continuity or contiguity of coast-line, is what I firmly believe. The line of migration of most of these shell-fish, was most probably from west to east, from America to Europe, during a different state of physical conditions from those which now prevail on our side of the ocean. But there has also been a march in the opposite direction, for we find some few littoral shell-fish (*Trochus cinerarius, Rissoa interrupta, Patella pellucida,* and the common cockle, *Cardium edule*), extending from the coasts of France to Finmark, but not reaching Greenland; whilst, as we shall see in our account of the Boreal province, others do not get so far. The common limpet (*Patella vulgata*) is said not to extend beyond Nordland, and the larger periwinkle (*Littorina littorea*), advances as far as Vadsoe.

These are small facts, but they have a large significance. The student of history follows, with intense interest, the march of a conqueror, or the migrations of a nation. The traveller traces with almost breathless delight, every step of the progress of some mighty hero of ancient days. I have had my share of the pleasure when tracking the course of Alexander and his armies in Pisidia, and determining mile by mile the route of Manlius through Milias; on ground, too, to the modern geographer, wholly new. Yet, absurd as it may seem to those who have not thought of such things before, there is a deeper interest in the march of a periwinkle, and the progress of a limpet. It is easier to understand how the son of Philip made his way safely through the sea, on his famous march from Phaselis, than to comprehend how the larva of a Patella crossed the fathomless gulf between Finmark and Greenland. It is a strong saying, but not said without a meaning, that the existence of Alexander may have been determined by the migration of the shellfish. If I am right in my interpretation of the reason why we find the same species of periwinkle in Greenland, and along the coast of Labrador, that lives now also on the shores of Nordland and Finmark,—in the unravelling of the cause and means of its wanderings, we acquire a clue to the origin of the peculiar physical conformation of the world as it is, and to the disposition of those geographical arrangements upon which the development of nations and characters of men in a great measure depend.

CHAPTER III.

BOREAL PROVINCE.

THERE is something in the atmosphere of northern regions that makes men worshippers of Nature, unattractive as is her boreal aspect during no small portion of the year. Whilst the short but genial summer lasts, her charms, however, as if in compensation, burst forth with multiplied attractions, and the torpidity of the observing faculties, whilst the long winter is dragging her not unpleasant course, seems to give double force to their powers, when, waking from their coerced sleep, they are attracted by the thousand objects glowing into life and beauty on every side around us. In the pleasant regions of the south, where all seasons teem with creatures "fair to see," and beings curious to observe, men, living continually amid scenes replete with beauty, are content to let their sensual perceptions overcome intellectual efforts, and amidst a continuous profusion of objects are happy to delight in their presence, and to revel amid the charms of creation, without making an effort to investigate the nature of the things that constitute the elements of these charms. Unremitting over-abundance has a like effect on the intellectual and the physical energies of man; it depresses and overpowers, and instead of profusion being a blessing, it is too often a curse upon his exertions. Thus, in the wide expanse of South America, we find the regions where vegetation exhibits a luxuriance, and the soil a richness beyond that of every other country in the world, where the earth and the waters alike teem with food, man, whether the aboriginal savage or the invading settler, sinks into an animal, who makes no effort towards improvement, and takes no thought of the morrow. But in the most dreary, and unpromising districts of the same great continent, the very cheerlessness and absence of attractions and comforts generate energy and success in the inhabitants. To some comparable influence on man's mind may we not attribute the intellectual energy of northern men as compared with southern, and the superior acuteness of their observing powers, and consequently of their abilities and knowledge as naturalists? The mould in which the character of a nation is cast, is like most moulds, a mineral one,—the soil and its properties,—and the power which melts the metal, and shapes it to the mould, is the influence of temperature, whether it be a man cast by God, or a spoon cast by man. The sun and the earth, climate and soil, are the great ethnogenitors.

To the pursuit of knowledge under difficulties may fairly be attributed the bias of Scandinavian minds towards the study of Nature, in all her aspects, and to the investigation, so energetically pursued, of the Fauna and Flora of their region. The naturalists of the Boreal regions are almost always intimately acquainted with the creatures of the countries in which they live; a remark which cannot be applied to all parts of Europe. Linnæus, that mighty mind, who, unquestionably, by systematizing the entire length and breadth of his science, laid the foundations of the vast superstructure which is fast attaining majestic dimensions, set the admirable example of investigating, in all its details, the natural history of his own country. And ever since his time, the naturalists of Scandinavia have been indefatigable in the exploration of their native lands and neighbouring seas. To mention those who have worked with success the marine natural history of the Boreal province, would be to fill pages with long arrays of eminent names, a catalogue not likely to be suddenly terminated, since the same spirit is at work in the north, and new candidates for fame are yearly appearing.

I must content myself by referring to a very few among those now living, whose writings more especially concern the subjects of this chapter,—to the veteran Nilson, to Löven, to Sars, Steenstrup, Eschricht, Kroyer, and Oersted. One name among the many illustrious dead, that of Otho Frederic Müller must not be unrecorded in any work on the "Natural History of the Sea."

The Boreal province may be regarded as the

meeting and mingling ground of the Arctic and Celtic faunas. Professor Lövén * remarks that the Scandinavian Seas "belong to two different regions, in the south the Germanic, in the north the Arctic. The fauna of the German Ocean prevails from the Sound to about the promontory of Stadt, on the coast of Norway, and decreases from thence to the Westfiord (Loffoden Islands) north of which it is subordinate to the Arctic fauna, which predominates from the North Cape to the Mitfiord, mingles with the Germanic to about Bergen, and decreases south of that point till it reaches its minimum on the coast of Bohuslan (Sweden). The character of the Germanic fauna is European, that of the Arctic, Atlantic. But a time was," continues my valued correspondent, "when the Arctic fauna extended over the whole of our peninsula down to its southern parts, as is proved by the fossils in raised sea-beds and pliocene strata, which, in places near the actual sea, are of species now living there; but further inland, of species now existing in the northernmost parts of Scandinavia, or only in the seas of Spitzbergen and Greenland, or even in some few cases, perhaps, extinct, and at these localities all the more southern species of the present German Ocean are wanting. Now this ancient Arctic fauna indicates an Arctic climate over the whole of Scandinavia. It is probable, that the currents of the sea during that period, were Polar currents, with a general

* In Letter, January 1847.

direction from north to south, and that since then their direction has been changed to that now prevailing, from south to north. The consequence of this was the migration of southern (Mediterranean) species northward, until they reached our coasts, and of the original Arctic species also to the northward, till some of them were actually driven from the 'land of their fathers' to the cold seas of Spitzbergen—migrations that are going on, perhaps, at this day, though of course very slowly. But before the Mediterranean species arrived, our shores were peopled with a number of species probably from the Celtic regions, which, being at present neither Mediterranean nor Arctic, and obtaining in the Germanic region their maximum of development, appear to have finally settled in that sea, *quibus mare Germanicum germana patria*. So I get in each of my two regions, *Regio Germanica* and *Regio Arctica*, three tribes, *Hospites e mari Siculo, cives Germani* and *aborigines.*"

Some years must elapse before we can determine the category (according to the ingenious distinctions suggested by Professor Löven), to which each animal form (and vegetable also) should be referred. When a stranger species of prolific habits, and capable of adapting itself readily to the new conditions under which it is placed, has colonized an area for some time, it is exceedingly difficult to distinguish between it and a true aboriginal.

Throughout the Boreal region Cetacea are abundant, the smaller whales especially, and every here and there, becoming more and more frequent as we proceed northwards, great "finners" may be seen, the giants of the ocean, steaming rapidly across the deep, and, in spite of their enormous bulk, rivalling vivacious porpoises in their gambols, as I once witnessed even as far south as the Zetlands. But the great Greenland whale is wholly absent from the Boreal province, a negative character of no small importance. Some of the dolphins are found here in wonderful abundance, especially the bottle-nosed whale, shoals of which occasionally strand themselves on the islands and the mainland of this region, bringing a rich harvest of oil and blubber to the fortunate fishermen, in whose neighbourhood they make their luckless landing. The numbers of the caa'ing whale (*Delphinus melas*) cast ashore in Faroe in 1843, according to Sir Walter Trevelyan, was 3146, from which 87,404 gallons of oil, the value of which was 5665*l*. were obtained ; the flesh, moreover, was cut into long strips, and dried for the purpose of feeding cows, who throve upon this novel food, and produced very excellent cream. The lives of no fewer than 600 cows were calculated to have been saved in one winter by this means.

The Boreal region is well characterised by its more peculiar fishes, especially those inhabiting the deeper parts of the Norwegian seas. Nowhere in Europe are fisheries habitually conducted at such great depths. Roving among groves of gigantic zoophytes, at a depth of one hundred fathoms or more, where the bottom is rocky, lives the red-fish, *Sebastes Norvegicus*, a sea-perch much sought after for food, and caught by the hook. Along with it are the *Macrurus Norvegicus*, and the "King of the Sea," as the Norwegian fishermen style him, the *Chimæra monstrosa*, grotesque and ferocious in habit; also, strange as it may seem, the *Coregonus silus*, a fish of the salmon tribe, belonging to a genus of which almost all the species are confined to fresh water, whilst this one is a dweller in the deepest and saltest parts of the habitable ocean. A curious shark, the *Spinax niger*, remarkable for the glistening aspect of its rough skin, which, when seen fresh from the water, appears as if frosted with needles of glass, is another citizen of these abysses. In great depths—as much as two hundred fathoms—not far from shore and never far out at sea, lives the *Lota abyssorum*, a fish of the cod tribe, not found southwards of this province. Members of the cod tribe are, indeed, very characteristic of this Boreal region; the ling, the tusk, the various kinds of *Merlucius, Pollachius, Merlangus* and *Gadus*, give a *facies* to its ichthyology. The *Brosmius vulgaris*, or tusk, is especially representative of this fauna, extending its range from the Zetland seas to the Polar circle. It is an excellent fish for the table, as I have experienced; it has a lobstery consistence and flavour, which tastes on the palate as fish and sauce conjoined. Epicures should make voyages to the north to eat tusk. The ling (*Lota molva*) and it are both dwellers in the deep sea, usually far from land, and the pursuit of these fishes employs thousands of fishermen, whose adventures are most perilous, and whose lives often fall sacrifices to their scantily-rewarded toil. The true cod, the hake, and the coal-fish frequent most the region between fifty and fifteen fathoms, the cod preferring the lower part of this region, the hake the upper. In shallower depths, and to the verge of the shore, the pollack prevails, and takes the place of its congeners. Mingled with these, on the Norwegian shores, is the green cod, or sei, the *Pollachius virens*, which, however, is more characteristic of the southernmost portion of the Arctic province, where it furnishes abundant employ to the fishermen of Finmark and Nordland. In this nursery of Boreal fishes, we must not forget that the herring and the halibut have their share in these northern seas.

To the clergyman of a remote country parish, in the wildest part of Norway, we are indebted for our knowledge of the more remarkable marine animals of the Bergenstift, or district of which the prettily-situated and flourishing town of Bergen is the capital. More complete or more valuable zoological researches than those of Sars, have rarely been contributed to the science of Natural History, and the success with which he has prosecuted investigations claiming not only a high systematic value, but also a deep physiological import, is a wonderful evidence of the abundance of intellectual resources which genius can develope, however secluded and wherever its lot be cast. How many an involuntary recluse, in far more favoured climes, drags heavily his time as if it were a chain, and bemoans piteously his hard fate in being shut out from all community with the intellectual world and the objects of its studies. Let him take a lesson from the course of the Norwegian priest, and learn that everywhere there is employment for the active mind, sources of continual enjoyment and instruction; that, in the most lonely places, God's book of nature lies open on the mountain and by the sea-side, with many a page in it unscanned as yet by mortal eye, and many a new and wondrous history as yet unperused. Even if the excitement of fame be sought for, it is not forbidden; the name of Sars, who reaped reputation when seeking no more than knowledge, familiar to every naturalist in Europe and America, in Asia, and at the Antipodes—for there are great naturalists settled far in the south, and many in the far east—is a sufficient proof that able work brings the rewards of applause and veneration, even when they be unasked for.

Sars has especially directed attention to the distribution of marine animals and plants on the coasts of the province in which he has fixed his habitation. In the tract from high-water mark down to the great sea-weed belt, he recognizes four regions. The highest of these is the region of Balani, where the barnacles grow in such numbers on the rock-side as to belt the coast, when the tide is out, with a white girdle. The second is the region of Limpets, at the upper bounds of which the sea-weeds *Fucus vesiculosus* and *Fucus nodosus* grow, and, lower down, *Fucus serratus* and *siliquosus;* in this region are seen numerous littoral shell-fish, species of *Littorina, Patella vulgata*, and, less plentifully, the tortoiseshell limpets (*Acmæa testudinaria*), *Purpura lapillus* and *Mytilus edulis*, the common mussel. Also, shell-framing annellides of the genus *Spirorbis* and red *Actineæ*, probably *A. mesembryanthemum*. Many Gasteropoda and *Ascidiæ* live here. His third belt is the region of Corallines, meaning by that term the pretty calciferous sea-plant, *Corallina officinalis*. This is the home of the horse-mussels (*Modiola modiolus*), of the large and showy *Actinia coriacea*, of *Lucernariæ*, Ascidians, sponges, and Alcyoniums. In sandy portions of this region numerous soft worms live (*Arenicola, Nephtys, Terebellum, Cirratulus* and *Aricia*), and, burying in the sand, we have bivalve shell-fish, of the genera *Mya* and *Solen*—to use their popular names, gapers and razor-fishes. This is the home of *Ciona intestinalis* and *Eolidia papillosa*. The fourth, and lowest of these coast-line belts, is the region of Laminariæ, of the great sea-flags or tangles, which lies beyond the lowest ebb. On the frond of these sea-shrubs live numerous and beautiful species and genera of Nudibranch mollusca and other *Gasteropoda*, the blue-dotted limpet, *Patella pellucida*, star-fishes, many *Actineæ* and numerous species of *Caprella* and *Nymphon;* and, on their sturdy stems are assembled *Ascidians, Alcyonia, Tubulariæ,* corallines and *Ophiuræ*. Sea-urchins, and the larger forms of star-fishes, including *Goniaster equestris*— so rare to the south of the Boreal province—live on the rocks. Beyond the boundary of this Laminarian region, multitudes of invertebrata reside. With the details of their distribution on the Norwegian coasts we are, however, insufficiently acquainted, but have reason to believe that they do not differ materially from the arrangements presented by the similar animals in the sea around the Zetland Isles, where I have personally investigated them.

The observations of Professor Löven,[*] on the Bathymetrical distribution of submarine life in the Scandinavian seas, bear out those of Sars, and carry our knowledge of them into the depths. "The littoral and Laminarian zones," he states, " are very well defined everywhere, and their characteristic species do not spread very far out of them. The same is the case with the region of frondaceous Algæ, which is most developed nearer to the open sea. But it is not so with the regions from fifteen to one hundred fathoms. Here there are at the same time the greatest number of species and the greatest variety of their local

[*] Brit. Assoc. Rep. vol. xiii.

assemblages; and it appears to me that their distribution is regulated not only by depths, currents, &c., but by the nature of the bottom itself, the mixture of clay, mud, pebbles, &c. Thus, for instance, the same species of *Amphidesma* (*i. e. Syndosmya*), *Nucula*, *Natica*, *Eulima*, *Dentalium*, &c., which are characteristic of a certain muddy ground at fifteen to twenty fathoms, are found together at eighty to one hundred fathoms. Hence it appears that the species in this region have generally a wider vertical range than the littoral, Laminarian, and perhaps as great as the deep-sea coral. The last-named region is with us characterized in the south by *Oculina ramea* and *Terebratula*, and in the north by *Astrophyton*, *Cidaris*, and *Spatangus purpureus* of immense size, all living, besides *Gorgoniæ* and the gigantic *Alcyonium arboreum*, which continues as far down as any fisherman's line can be sunk. As to the point where animal life ceases, it must be somewhere, but with us it is unknown. As the vegetation ceases at a line far above the deepest regions of animal life, of course the zoophagous mollusca are altogether predominant in these parts, while the phytophagous are more peculiar to the upper regions. The observation of Professor E. Forbes that British species are found in the Mediterranean, but only at greater depths, corresponds exactly with what has occurred to me. In Bohauslan (between Gottenburg and Norway), we find, at eighty fathoms, species which, in Finmark (on the north), may be readily collected at twenty, and on the last-named coast, some species even ascend into the littoral region, which, with us here in the south, keep within ten to eleven fathoms."

The great tree *Alcyonium*, a branched zoophyte of leathery texture, alluded to by Professor Löven, is a very wonderful and characteristic production of the abysses of the Boreal seas. The lines of the fisherman, when fishing for the red-fish, or uër, become entangled in its branches, and draw up fragments of considerable dimensions, so large, indeed, that the people of the country believe it to grow to the size of forest-trees, an exaggeration, in all probability, but nevertheless one founded in unusual magnitude. It appears to me that many of the bodies to which geologists have given the name of *Fucoids*, and too hastily assumed to be plants, were creatures allied to these *Alcyonia*, and, possibly, some of them to *Alcyonidium*, a similar body of a different class. This notion of their nature is much more consistent with the character of the strata in which they occur and that of the fossil fauna with which they are occasionally associated. The number and beauty of kinds of sea-urchins, star-fishes, and sea-cucumbers, give a characteristic feature to the Boreal seas, which, in this respect, are more prolific than the Celtic, and probably also than the Mediterranean province. It is true that we count as many species in our British lists, but then a portion—and no inconsiderable one—of the array is derived from that part of Britain which falls within the Boreal area.

The Norwegian Echinodermata have been made the subject of an excellent monograph by Von Duben and Koren.* They enumerate two species of Crinoids ; three of *Euryales*, a group especially characteristic of this region in the Atlantic ; ten of *Ophiuridæ*, one of which is not known to the south of Norway ; eighteen of *Asteriadæ*, including a peculiar *Solaster*, and species of *Astragonium*, *Pteraster*, and *Ctenodiscus*, in all seven, not known as Celtic forms ; thirteen sea-urchins, two of them confined to the Norwegian seas, and fourteen sea-cucumbers, of which three are not known out of Norway. The majority of species are distributed all along the coast of Norway, both south and west ; but several forms common to the Arctic province, occur in the latter district only. The crinoids, the species of *Astrophyton* (or *Euryale*), the *Cidaris papillata*, and *Brissus fragilis*, are remarkably characteristic of the region of one hundred and more fathoms. It is worthy of note that extreme brilliancy of colour is exhibited by the Boreal Echinodermata. For vividness of painting, and elegance and variety of pattern, few marine animals can equal the northern brittle-stars. The cushion-star is of the most dazzling vermilion ; and almost every kind of star-fish and sea-urchin displays gorgeous contrasts of red, blue, green, purple, and yellow.

* Vide Kongl. Vetenskaps-Akad. Handlingar, 1844.

The distribution of submarine creatures in the fiord of Christiana at the south-eastern angle of Norway has been inquired into by Örsted, and the result of his researches* shows that the features of the Boreal province are there slightly modified by the Celtic fauna. Such mollusks as *Chiton marmoreus, Nucula tenuis, Syndosmya intermedia, Cemoria noachina,* and *Astarte elliptica,* accompanied by the Echinoderms, *Echinus neglectus, Goniaster granularis, Brissus lyrifer, Cuviera squamata,* and *Holothuria elegans,* determine, however, the strict connection of the fauna of the southern shores of Norway with that of her western coast. From the coast-line downwards there appear the usual sequence of green, brown, and red Algæ, and the depths are characterised by the beautiful coral *Oculina prolifera.* The *Goniaster* above noted ranges between thirty and sixty fathoms, and the *Holothuria* lives at a depth of eighty fathoms. The worm-inhabited tooth-shell, *Ditrupa,* occurs in fifty fathoms water, with the sea-rod, *Virgularia,* which ranges to sixty fathoms; the animal-flower, *Anthea cereus,* to eighty fathoms; and that most curious of sponges, the *Tethya cranium,* so like an infant's head that we might almost fancy it the capital extremity of some new-born merman-child whom our dredge had decapitated in its mother's arms, is found there, as in Zetland, at a depth of eighty fathoms.

Iceland sharing in the features of the Arctic and

* See Kroyer's Tidoskrift, for 1845.

Boreal provinces, and constituting, so far as Europe is concerned, the westernmost boundary of both, appears to present a fauna which is very closely comparable with that of Finmark and Nordland. The vaagmer, the tusk, the abundance of cat-fish and lump-fish, the presence of herrings in considerable shoals, and of ling, skate, and halibut, is an assemblage which gives a truly Boreal character to its ichthyology, whilst the visits of the capelin show how it passes into the Arctic province, further indicated by the visits, few and far between, of the Greenland whale. Fin-fish, bottle-nosed porpoises, and seals, including the *P. barbata, leporina,* and *groenlandica,* show a similar rule among the marine mammalia. A full account of its marine invertebrata is a desideratum which we may look to the able naturalists of Denmark to supply. Sir William Hooker was struck with the scarcity of shells on the Iceland shores; among the few he saw were the *Mya truncata* and *Venus islandica.* Judging from the list of Iceland sea-weeds given in the account of the voyage of the "Recherche," there is, however, in all probability a considerable population of Mollusca, Crustacea, and Annellida, inhabiting the Laminarian zone.

The natural history of the Zetland Islands clearly indicates their position within the Boreal province, and their marine zoology is conspicuously of the Norwegian type. This group of bare and barren islands, so bare that the unique tree, some ten or twelve feet high, is shown as a curiosity; and so barren, that the unproductiveness of the soil produces more famines than food for the people, offers but few attractions to the terrestrial zoologist, or to the botanist; their birds, chiefly of remarkable northern types, and their one peculiar plant, the pretty little *Arenaria Norvegica*, excepted. But the deficiency in the land is fully compensated for by the redundancy in the waters; and in no part of the British islands is the naturalist so sure of reaping a rich harvest as in the Zetland seas. The coast-line, the bays or *voes*, and the deep sea or *haaf*, equally abound in singular and interesting forms of Boreal life. The tides have but a small fall; yet between high and low-water mark an ample harvest of curious creatures and marine plants may be gathered. In the Laminarian zone the great roots of the tangles are inhabited by thousands of creatures, specifically new to the zoologist who comes here from the southern shores of Britain. *Margarita undulata* and *Trichotropis borealis*, appearing in numbers, soon inform him of his latitude. But above all, the quantity of *Holothuriæ*, sea-pudding as the natives call them, attracts and astonishes the dredger. The great *Cucumaria frondosa*, whose body, resembling a huge sausage, when extended reaches a length of three feet, occurs in abundance and furnishes admirable subjects for the skill of the anatomist. The sheltered bays swarm with medusæ; many of these kinds not seen elsewhere in the British seas; countless shoals of the curious little *Lizzia octopunctata*, with its jet black eyes; swarms of *Thaumantias pilosella*, like so many coronets circled with rubies; *Circe rosea*, the most elegant of submarine mitres; and *Steenstrupia rubra*, jerking itself in all directions, trawling its single tentacle after it, as if it were attacked by some ferocious vermilion worm, mingled with the graceful *Briaria*, the swift *Sagitta*, and iridescent crowds of *Mnemiæ*, *Beroe*, and *Cydippe*, give a distinctive character to these our northernmost British waters. When the dredge is plunged into the depths, whether near or far from shore, it comes up filled with Norwegian animals. *Echinus neglectus* in the shallower localities, *Echinus Norvegicus* in the deeper, especially distinguish the region, and deeper still there is the rare and beautiful *Cidaris*, whose long and slender spines have suggested the local name of "piper." With it is associated the true Medusa's hand, that strange star-fish with arborescent arms, known scientifically as the *Astrophyton* or *Euryale*. The rude yet not unintelligent fishermen are attracted by the curious creatures which cling to their lines when they are engaged in the perilous occupation of fishing for the ling, itself a characteristic feature of these seas, on deep-sea banks, some twenty or thirty miles from shore, far out in the clear ocean, whence occasionally resisting their superstitious prejudices they bring to the shore specimens worthy of national museums. One of their favourites is the "sea-apple" (*Tethya cranium*), a curious globular sponge of a bright sulphur-yellow colour, and as large as an orange. Occasionally they bring up the great Madrepore, striking among Boreal productions. A rare but excellent fish is the tusk, and another of their curiosities is derived from among the vertebrata, being that extraordinary shark, the *Chimera monstrosa*. These seas are frequented by the lesser cetacea, and not unfrequently by finner whales of considerable dimensions. The Zetland seas were the scene of the earlier researches of Professor Jameson, and of Dr. Fleming, names that will ever shed a lustre on British natural history. The seals which frequent them (the great *Phoca barbata* is one) have been carefully studied by Dr. Edmonston, himself a Zetlander, and whose most promising son, the author of a "Flora of Zetland," held out hopes of high scientific distinction, alas, prematurely arrested by his accidental death when prosecuting his researches on the coast of Peru. Of late years, Mr. M'Andrew has cruised with great success in this interesting district, cruises in which I have had the great pleasure of sharing, and of aiding in gathering an abundant store of valuable observations in all departments of our science.

CHAPTER IV.

CELTIC PROVINCE.

The Celtic province is our home-circuit. Above all other maritime regions it has been chosen by naturalists for their minutest observations. If their science, so far as it concerns the sea, was born, as some have said, in the Mediterranean, it was brought up in the British Channel, and on the mid-western coasts of the Continent. From the bay of Biscay to the Baltic sea, there has been and continues a diligent and searching investigation into the nature and species of the animals and vegetables that live beneath the waters. Their abundance, and the facility with which they can be procured, have been main causes of the attention devoted to them. But, however plentiful, or however easily procurable, we should have learned comparatively little about them had the spirit of energetic research and minute enquiry, characteristic of the enlightened portion of the human population of these regions, been absent. In the British Islands, Natural History has long been a favourite pursuit; one indigenous, in a manner, to the people, and attractive to them for its own sake. It leads to no profit, no high places, no honours, no social position; it has no academical distinctions accorded to it, and the few official posts connected with the study of it are but poorly remunerated and unattractive. Nevertheless, the number of naturalists, of one grade or another, is very considerable, and greater in Britain than in any other civilized country. The majority are men highly enlightened and of a liberal and far-seeing spirit. They are to be found in all classes of the community; mostly in the middle ranks; not unfrequently among the lower classes, and sometimes, though unfortunately but seldom, among the aristocracy; this is the more to be regretted, since for men with cultivated minds, and abundant leisure and wealth, the study of Natural History is peculiarly adapted. The neglect of this science in our universities is the cause of the defect. Sooner or later, it will be remedied, when its unquestionable educational value shall be turned to account. It is in vain that we erect museums and amass valuable and extensive collections, if we discourage the acquirement of the knowledge for the illustration of which all this scientific display is prepared. We boast of our vast cabinets of objects of Natural History, and, in the same breath, question the propriety of teaching men the meaning of these treasures of divine workmanship. We complain of the want of teachers, yet make but unwilling efforts towards training students for the duty of instructing others. If there be one land above all other lands favoured for the study of Nature under all her various avatars, it is the goodly island that Providence has, in favour, given us for a birth-place and home. If there be one region above all other regions fitted to be constituted the type and model, whether through the variety of its inhabitants, their abundance, or their convenient collocation, it is the Celtic province of which the British Islands seem to constitute the centre.

The Celtic province is the neutral ground of the European seas; it is the field upon which the creatures of the north and those of the south meet and intermingle. It has its own special inhabitants, the aborigines of the province, but these are far exceeded in numbers by the colonists who are diffused among them. It includes within its proper population the survivors of an epoch when the seas of Europe were differently parcelled out than they are now. Here and there, these old people still retain limited tracks of the sea-bed, whilst the vast mass of the nations to which they originally belonged have retired far to the north, or west, or south, according to their tribe. These must not be confounded with the immigrants who have gradually made their way into the Celtic area during the ages that have past since its first constitution into a distinct province. They are like the Basques among the Spaniards, or the Cornish among Englishmen, relics of ancient possessors of the country whose epoch of dominance has ceased to be, but who still remain in fragmentary masses, as if to show what and where they once were. These varied natural-history features, combined in the Celtic province, render it of all European areas that most interesting to the zoologist and botanist; from their abundance and interest, they incite the human inhabitants to the study of the living creatures gathered so profusely around them : hence it is, that, in spite of all the discouragement just alluded to, in no part of the world has marine natural history been so thoroughly pursued as in Britain.

The area of the Celtic region has its southern limits about Cape Finisterre, and at the entrance of the English Channel. All the German Ocean, with the exception of a belt skirting the southern coasts of Norway, may be said to belong to it, and all the seas immediately around the British islands, excepting about Zetland. The Baltic appears to be an arm or extension of it, carrying its fauna far to the north of its normal limits. A great part of this region is comparatively shallow. Very deep water (depths below the hundred-fathom line) approaches nearly the western coasts of Ireland and Scotland. These abyssal gulfs probably limit this extension of the characteristic Celtic fauna. Occasional tracts of very deep water, ravines, or pits, as it were, such as the line of deep below 100 fathoms, between Galloway and the opposite coasts of Ireland, here and there occur. In the instance mentioned, an insulated ravine, its sides from 60 to 80 fathoms below the surface of the sea, and its bottom 150 fathoms deep, extends for 30 miles, with a breadth of not more than 2 miles.

The floor of the Celtic province may be regarded as an elevated platform with steep sides, deep isolated pits and furrows, indenting bays and gulfs. In its southern and western divisions this submarine table-land supports a numerous population, but that section of it constituting the bed of the North Sea is comparatively thinly inhabited. The deep parts of this latter portion, however, swarm with fish and other animals. The little silver pit, 330 feet deep, may be cited as an instance. The line of 100 fathoms may be taken as the southern Celtic boundary. It pursues its course wavily and with a general outward curve from off the coast of Kerry to near the northern extremity of the Bay of Biscay. The fifty-fathom line runs from Scilly towards Ushant, with a deep inward sinuosity, and between Scilly and the southernmost coast of Ireland makes a profound bend up St. George's Channel. The shallows of the inner extremity of the English Channel are impediments to the spread of many species.

To the physical phenomena of the Celtic area, and the geological changes it has undergone, are due those varied features which its fauna and flora present: warm currents from the south, cold currents from the north, coast-currents, and oceanic currents, all converge to it as a centre. In their course migrate fishes, crustacea and mollusks slowly but surely, and migrations, conducted during a long series of ages, have mingled together the creatures of many climes and regions. The ancient meeting-place of glacial and warmly-temperate seas, as successive geological events changed the orography of the land, and the hydrography of the ocean, the animals dwelling side by side under those opposite climatal conditions, did not wholly disappear, but remained in part to bear living witness of their ancient extension. All the changes within the Celtic area have been beneficial, and to the establishment of a Celtic and strictly-temperate province in the interval made by the recession of opposing climates, its richness at present in organized treasures is mainly due; for on the events which brought out such a result, depended the peculiar arrangement of currents, such as we now find around the coasts of Britain, which, by their constant action, have had so powerful a share in determining the natural history of the British seas.

Along the coast of Belgium and Holland, and on to the low and sandy shores of Denmark, the marine fauna and flora are scant and poor. Tracts of sand, when of great extent, are unfavourable to the spread and variety of aquatic forms of life, even as they are obnoxious to terrestrial creatures. In a confined and sheltered space, such as the strait between Denmark and Sweden, however, there is a more abundant development of the population of the sea, even though its extent be limited by the deleterious influence of the brackish waters that flow from the Baltic. An excellent account of the conditions and phenomena of submarine life in the Strait of Oresund, has been published by A. S. Orsted.* This essay should be studied by every naturalist interested in such inquiries. In the locality explored, this able observer distinguishes three regions of submarine vegetation. The first is that of green sea-weed, REGIO CHLOROSPERMEARUM, It extends from the highest sea-mark to a depth of from 2 to 5 fathoms. Its upper portion is the sub-region of *Oscillatorineæ*, and is that part most frequently exposed to the air. Its lower portion is the sub-region of *Ulvaceæ*, where the sloke-plants, *Ulva lactuca* and *latissima*, with various *Confervæ* and species of *Hormiscia*, *Ulothrix*, and *Cruoria*, flourish. A few olivaceous algæ, and some purple ones, but never those that are of brilliant hues, also occur. The second region is that of the olive-coloured seaweed, REGIO MELANOSPERMEARUM, extending to 7 or 8 fathoms. It is constituted also of two sub-regions; the uppermost is that of *Fucoids* and *Zostera*. "This," remarks the describer, " is, as it were, the savannah of the sea, for the *Zostera marina*, which, here ruling, has so much of the aspect of a grass, that the fishermen call it sea-grass,

* " De Regionibus Marinis. Elementa Topographiæ Historico-Naturalis freti Oresund." Havniæ. 1844.

extends over a great space on the sea-bottom, with an uniformity comparable with that of a tropical savannah." On a stony sea-bed, as usual elsewhere, *Fuci* take its place. The lower sub-region is that of *Laminariæ*. "Hæc subregio silva maris haberi potest; *Laminariæ* enim, 10—15 pedes altæ, erectæ velut arbores silvæ, confertæ sunt." The third and lowermost region is that of the purple sea-weed, REGIO RHODOSPERMEARUM. Its proper range is from 8 to 20 fathoms: it is confluent with the last. Its characteristic algæ are *Iridea edulis, Delesseriæ, Hutchinsiæ, Callithamnia, Ceramii, Gigartinæ* and *Odonthalia dentata.* If this classification of zones of vegetable life be compared with the brief notice I have given of the subdivisions of the littoral and laminarian zones on the British coasts, a close correspondence will be perceived; indeed, the chief difference lies in the stress laid upon the relative value and connection of the sub-regions.

M. Oersted has given some interesting tables of the relations of the Algæ to light, sea-composition, and depth in the locality explored. These I abstract, in order to call attention to this most interesting subject, and because the memoir in which they are contained, is not likely to be within the reach of many British naturalists, having been published in the form of an inaugural thesis.

The first concerns colour, in its relation to depth.

COLOR.	ALGÆ.	PROFUNDITAS.
Radii violacei ,, cyanei ,, cœrulei	Algæ viridicœrules- centes (Oscillatorineæ)	Superficies
,, virides	Algæ virides (Chlorospermeæ)	Ped. 10—25
,, flavi ,, aurantiaci	Algæ olivaceæ (Melanospermeæ)	Ped. 25—50
,, rubri	Algæ purpureæ (Rhodospermeæ)	Ped. 50—65

The second exhibits the influence of sea-composition, intensity of light, and motion of the water upon the three groups of green, red, and olive seaweeds.

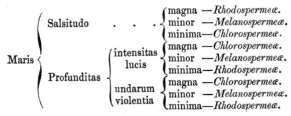

M. Oersted, after having described his vegetable zones, then proceeds to constitute three zones of animal life. The first is the REGIO TROCHOIDEORUM, ranging from the shore sea-mark to 7 or 8 fathoms. He remarks that the shells of the testacea in this province are strong, in order to endure the force of the waves; those that have no shells can hide or bury themselves. Its inhabitants are, for the most part, phytophagous animals. It may be separated into several sub-regions. The first and uppermost is that of *Littorinæ*, where these mollusks are associated with bivalves of the genus *Mya*, and with worms of the genera *Nereis*, *Spio*, and *Arenicola* (lugworms). The second is the sub-region of *Mytilus edulis*, the common mussel. It corresponds to the sea-wrack's domain. Here we find the *Akera bullata*, the *Ascidia intestinalis*, the common sea-urchin (*Echinus sphæra*), and the harry-crab (*Carcinus mænas*). The third sub-region is that of the little whelk, *Nassa reticulata*, which, with the small bivalve, *Corbula nucleus*, prevails here. Several fishes are dwellers here, as *Spinachea vulgaris*, *Cottus scorpius*, the gunnel, the viviparous blenny, the fluke, and the sand-eel. Certain shell-fish, normally freshwater species, intrude under Baltic conditions here, such as *Limnæus Balticus* (*i. e.* a var. of *L. pereger*), and *Neritina Baltica* (*i.e.* a var. of *N. fluviatilis*). The second animal province is the REGIO GYMNOBRANCHIORUM. It corresponds with the Laminarian and Rhodospermean plant belts. It is but partial in this locality. Its inhabitants are often remarkable for colour and variety. Among mollusks there are Limpets, Chitons, *Ascideæ* and nudibranchs; among crustacea, *Caprellidæ* and *Pycnogonidæ*. Most of its characteristic inhabitants are soft, or at least not strongly protected. The third, and lowest animal province, is the REGIO BUCCINOIDEARUM; occupying the deeper part of the straits, mostly on a muddy bottom. Its population are chiefly carnivorous, and apt to live immersed in mud. When covered by shells, these are not remarkable for thickness. Hermit-crabs and spider-crabs are here; sea-mice (*Aphrodite*); large sea-worms; whelks (*Buccinum undatum* and *Fusus antiquus*) in abundance with *Aporrhais*, screw-shells and tooth-shells, quantities of *Leda rostrata*, also *Cyprina islandica* and *Hiatella arctica*. Sea-pens are also found here.

The scanty fauna of the Baltic is too decidedly Celtic to be regarded either as belonging to a separate province or to the Boreal region. The number of fishes of this sea, enumerated by Nillson, is under thirty. Several of these only range far through the southern portion of the Baltic, and none among them is peculiar to it. The herring is remarkable for presenting some peculiarities, and that of the northern half has been distinguished from that found in the southern part. These have been regarded as distinct from *Clupea harengus* by some ichthyologists, who have designated the former *Clupea membras*, and the latter *Clupea Cimbrica*: they seem, however, to be only slight varieties, due to the influence of peculiarities in the degree of freshness of the waters. The sprat occurs. The gar-pike frequents the southern district, and the fresh-water pike occasionally takes to the brackish waters; an important fact, when we consider the wide circumpolar distribution of this fish. Two or three members of the cod tribe make their way to greater or less distances, *Gadus callarius* having the farthest range. Of flat fish, the *Pleuronectes platessa* thrives throughout; also the *P. limandus* and *P. maximus*. Both our Celtic sand-eels occur, and a pipe-fish or two. The mackerel rarely enters this sea; the nine-spined stickleback is common, and a fresh-water species descends into the sea. Gurnards are rare, gobies and bull-heads common. The gunnel, the viviparous blenny, the sea lamprey, and the sturgeon, make up the summary of Baltic fishes. The invertebrate inhabitants of this sea are as few in proportion as the vertebrata. This character of its zoology is strikingly seen when we regard its molluscous population. Were the area, in any respect, the centre of a peculiar fauna, we should expect to find indications of a special creation, manifested by the mollusks. Instead of this being the case, once we have passed some way within the Sound, these animals become exceedingly few, whether we regard the number of genera or of species. A single periwinkle, the *Littorina rudis*, and its minute ally, the *Rissoa* (*Hydrobia*) *ulvæ*, slightly modified, miserably represent the long list of Celtic gasteropoda. A few bivalves, such as *Tellina solidula* and *tenuis*, *Donax anatinum*, *Cardium edule*, *Mytilus edulis*, and *Mya arenaria*, constitute the Lamellibranchs of this sub-region. Some of these, *Tellina solidula* for example, become slightly modified through the influence of local conditions, and have been elevated by over-anxious patriots into distinct species, with the distinguishing epithet of *Baltica* or *Balticus*, but a little consideration and comparison prove, beyond a question, that these are the merest local varieties. High up in the Baltic, there is a tendency towards a mingling of such marine and fluviatile mollusks, as can be inured to brackish water. Hence we find common forms of *Limnæus*, as *L. palustris* and *pereger* (var. *Balticus*), *Planorbis* as *P. albus*, *Bythinia* and *Neritina*, enumerated as inhabitants of the Baltic sea.

In the Channel Islands, and on the French coast of this region, we have evidence of the influence of a southern element, manifested by various well-known forms of fishes and mollusks, which either do not visit the shores of Britain, or are but rare and occasional visitants. The tracing out the course of this element, especially of so much of it as is littoral, would be a task well worthy of the attention of an expert field-naturalist. We must look to some of our able neighbours in France for the undertaking of this investigation. Along her Atlantic shores, some excellent naturalists have been at work : indeed, the first impulse to the scientific investigation of the distribution of marine animals was given by French zoologists working amid the sea-fauna of their own country. I allude especially to the researches of Milne Edwards and his colleague, Audouin. In their work, entitled "Recherches pour servir à l'Histoire Naturelle du Littoral de la France," published in 1832, they give an account of the result of their observations on the bathymetrical distribution of marine creatures, chiefly made upon the coast between Granville and Cape Frehel. They distinguish four littoral belts : a first and highest, dry at ordinary tides, where, when the coast is rocky, barnacles can live, but, if it be sandy, few or no marine animals are found : a second, which, in rocky places, is marked by a population of periwinkles, limpets, *Purpura*, *Nassa*, and red actineæ; and where sandy, crustaceans of the genera *Talitrus* and *Orchestes*, and the worms *Terebella* and *Arenicola ;* when muddy, besides these, occur *Nephthis* and small siphunculi. The third zone is chiefly characterized by the presence of corallines, and is only uncovered at low tides ; mussels, limpets, &c., are found on its rocks, green *Actineas* and compound Ascidians ; in other places nudibranchs, sea-ears, *Polynoes, Serpulæ*, and *Planariæ;* sponges, lobulariæ, and Ascidiæ garnish the interstices of large stones; millions of small *Cerithia* and *Rissoa* live among the grass-wrack, on its softer ground, and cockles, razor-shells, and clams, bury themselves in its sandy mud. The fourth zone, exposed only during the lowest tides, presents tangle-covered rocks, often studded with star-fishes. This is the domain of *Patella pellucida;* peculiar crustaceans and mollusks of the genera *Bulla* and *Pandora* live among the fine sand. At a lower level, never uncovered, is a fifth region, inhabited by oysters, cap-limpets, scallops, many forms of crustacea, sea-mice, large star-fishes, and peculiar worms. And, deeper still, there appears to be another region, in which none of these animals are known to be found. The energetic and philosophical naturalists, who, recorded these phenomena of distribution, foresaw how important such studies would become eventually through their geological bearings. " La distinction des divers niveaux qu'habitent exclusivement, et quelquefois d'une manière fort tranchée, les animaux marins, nous a paru d'autant plus importante à faire ressortir, que cette étude, poursuivie avec quelques soins, peut être un jour d'un grand secours à la géologie, et jeter une vive lumière sur plusieurs théories fondamentales de cette science."

In a catalogue of the marine testacea of the department of Finisterre, by M. Collard des Cherres, published in the fourth volume of the "Transactions of the Linnæan society of Bordeaux," are some interesting indications of the southernmost limits of the Celtic province. The general assemblage of mollusks and radiate animals in this locality is decidedly Celtic. Interspersed, however, are a few well-marked southern forms, either Mediterranean or Lusitanian, which do not reach to the British Channel. Thus we find in this list the names of *Purpura hæmastoma*, the whelk, which takes the place of *Purpura lapillus*, more to the south, though, in this district, the two species are together; *Nassa neritea*, a curious little whelk, resembling a *Nerita* in shape, abundant on sandy shores in the

Mediterranean, creeping on the sand, and burying in it near low-water mark, often in company with *Donacilla Lamarckii*, also a Finisterre shell; *Arca barbata* and *Lima squamata*, Mediterranean bivalves that live among rocks close to the water's edge: *Triton nodiferum* and *cutaceum*, whelks of a genus that has no representatives elsewhere in the Celtic province, and *Trochus Laugieri*, a sublittoral shell, not noticed further to the north. It is worthy of remark that the greater number of these species are dwellers on the verge of low-water mark, either above or below it. Moreover, they are mostly rock-shells, so that their presence here, separated frequently as they are from their brethren by the sandy shores of the southern half of the Bay of Biscay, is an anomaly not easy, at least by ordinary causes, to be accounted for.

Everywhere around the British shores the subdivisions of the littoral zone are strikingly marked by both animals and plants, especially on the more rocky portions of the coast. It matters not how great or how small may be the fall of the tide; the several belts of the zone are equally well distinguished where there is a very small and where there is a very considerable fall. There are local differences, especially noticeable when we compare the eastern with the western provinces, or the extreme north with the extreme south; but in the main the belts or subordinate zones are characterized by the same species throughout. Thus, the highest of them, that on the very verge of continual air is distinguished by the abundant presence of the seaweed named *Fucus canaliculatus*, among whose roots may be found crowds of small varieties of the periwinkle, called *Littorina rudis*, especially those forms to which the epithets *patula* and *saxatilis* have been applied. These, indeed, range out of the water considerably and may be found adhering to the rocks many feet or several yards above high-water mark. On the South-Western, and most of the Western provinces it is accompanied by a neat little black periwinkle called *Littorina neritoides*, a species which has a wide spread in the world, but is everywhere to be found in similar localities. The second sub-region is marked by the abundance of a small dark rigid sea-weed, called *Lichina*, painting the rock sides as if with a dingy stripe. With it we find the larger forms of *Littorina rudis*, abundance of the common limpet (*Patella vulgata*), the common mussel (*Mytilus edulis*) and myriads of small seaside barnacles. On parts of the coast where the shore is steep and rocky, even perpendicular, this belt may be seen striping the sea-wall like a broad white band, as if the strong boundary were overgrown by some hoary lichen. When we approach and peer into the cause, we find the whiteness to be owing to the presence of the shells of myriads of barnacles, all of one species of the genus *Balanus*, crustaceans, but very unlike crabs. Among them, on the barer portions of the rock, are fast-adhering limpets (*Patella vulgata*). Where the shore shelves a little, and rocky ledges decline gradually into the sea, numerous creatures are found living in this sub-region. In such a locality the common mussel delights to live, moored by its byssal cable in the crevices of rocks or, still more numerously, often in great companies, anchored among masses of gravel, the pebbles of which are tied together by its silky filaments. The rocksides and the floors of transparent pools are here often thickly coated with a hard pale red crust. This is a nullipore, in reality a seaweed, though putting on the aspect of a coral. Not very long ago it was regarded even by naturalists as a zoophyte, and is fairly believed to be a coral even at present, by fishermen who draw up branching varieties of it in their nets, from depths that are never uncovered by the tide. The region of half-tide forms a third subdivision of the littoral zone, one exceedingly prolific in marine animals and plants. Here we find *Fucus articulatus*, with its graceful even-edged rich brown fronds growing in profusion, mingled occasionally with the less elegant *Fucus nodosus*. Here limpets throng, and dog-periwinkles (*Purpura lapillus*), crawl observantly, seeking to bore more passive mollusks, and extract their juicy substance. This is the home of the best of periwinkles, the large black *Littorina littorea*, gathered in thousands for the London market. On our western coasts we find it in company with the purple-striped top shell

(*Trochus umbilicatus*), and towards the south with the larger *Trochus crassus*. Here are sea-anemones, especially *Actinea mesembryanthemum*, like masses of brilliant crimson or bright green pulp, but when covered by the water, expanding into many-armed disks, and displaying shapes and colours of exquisite beauty. A fourth sub-region succeeds, the lowest belt above low-water mark, distinguished by the presence of *Fucus serratus*, the saw-toothed shining black sea-weed, so much used in the packing of lobsters for market. It takes the place of *Fucus articulatus*. On its fronds creeps the lowermost in succession of the periwinkles, the variously tinted *Littorina neritoides*, exhibiting every colour in its obtuse and thickened shell, pure yellow, bright red, rich brown, dark olive, and all possible changes of striping and mottling. With it is associated everywhere *Trochus cinerarius*, except, and this exception applies generally to all the creatures whether animal or vegetable, where the coast is composed entirely of fine sand or clean gravel.

At the verge of low-water mark, immediately below it, wherever the coast is rocky, there are all round the British shores, within a space of a few inches, a remarkable series of more or less distinctly defined belts, each consisting of a different species of seaweed. These, in succession, are the *Laurencia pinnatifida*, uppermost; then the green *Conferva rupestris;* then the elegant and firm, often iridescent fronds of *Chondrus crispus;* and, lowermost, the thong-weed or *Himanthalia lorea*. Even when the others are absent, the last is usually present. Beneath all these, and extending to several fathoms deep, are the great *Laminaria* or tangle-forests, or, on sandy places, the waving meadows of *Zostera* or grass-wrack. Everywhere among the tangles, in the Celtic region, we find species of the periwinkle called *Lacuna*, and of the Limpet known as *Patella pellucida*, remarkable for its horny texture and translucency, and for the radiating rows of opaque spots of turquoise-blue decorating its surface. Here, too, are innumerable little univalve shells of the genus *Rissoa*, wonderfully varied in sculpture, colouring, and outline. This is the chosen haunt of the nudibranchiate mollusks, animals of exceeding delicate texture, extraordinary shapes, elegance of organs, and vividness of painting. Their bodies exhibit hues of a brilliancy and intensity such as can match the most gorgeous setting of a painter's palette. Vermilion red, intense crimson, pale rose, golden yellow, luscious orange, rich purple, the deepest and the brightest blues, even vivid greens, and densest blacks are common tints, separate or combined, disposed in infinite varieties of elegant patterns, in this singular tribe. Our handsomest fishes are congregated here, the wrasses especially, some of which are truly gorgeous in their painting. Here are gobies and more curious blennies, swimming playfully among these submarine groves. Strange worms crawl, serpent-like, about their roots, and formidable crustacea are the wild beasts who prowl amid their intricacies. The old stalks and the surfaces of the rocky or stony ground on which they usually grow are incrusted, like the trunks of ancient trees or faces of barren rocks, with lichenous investments. But whereas in the air these living crusts are chiefly, if not all, of vegetable origin, in the sea they are more often constructed out of animal organisms. Some of them are sponges, compound animals of the very lowest types; others are true zoophytes, polypes of simple structure, but often combined in complicated communities; others—perhaps a majority—resemble true corallines in general aspect, but differ importantly in essential nature, being *Polyzoa* or *Bryozoa*, beings that have proved to belong to the class of mollusca, however unlike they may seem to shell-fish. A *Flustra*, for example, is really a commonwealth of shell-fish, exceedingly minute, but each citizen, if we would compare it with the animal to which it has most affinity, is an inferior kind of *Terebratula*, or *Crania*. Each community is the result of the budding of some one individual, and wonderful indeed is it to contemplate the exquisite and defined beauty of each separate being, and the equally wonderful and regular conformation of the entire assemblage composing a single mass.

In the middle and lower part of the *Laminarian* region around our shores, the tangles become less plentiful as we descend, and at last become exceptional and disappear. But other sea-weeds are very abundant, especially those that delight in red or purple hues.

Sea-Vegetables of the Dulse tribe and its allies are very plentiful here, species of *Delesseria, Rhodomenia, Rhodomela*, &c. Tender sea-mosses, exquisitely delicate in form and colouring, species of *Hutchinsia, Callithamnium,* and *Ceramium* abound. Where none of these are very plentiful, we often find the coral-weed or Nullipore, in vast quantities, and assuming many strange modifications of form, growing in some places into miniature cabbage-heads or fucus-like expansions, in others assuming a truly coral-like aspect and deserving its popular designation. Among these vegetable corals numbers of peculiar fish, shells, and articulate animals delight to live; and probably not a few derive subsistence from their stony fronds. The *Lima* (a shell-fish related to the Scallop) gathers the broken branches by means of prehensile tentacles, and constructs for itself a comfortable nest, lined with a woven cloth of byssal threads. Numerous fishes resort to these rugged pastures in order to deposit their spawn among the gnarled branchlets.

The destruction of a nullipore ground is sure to drive away its finny frequenters, and, consequently, enactments have been made at various times in statutes concerning fisheries, for the preservation of this valuable variety of sea-bed.

The zoological and botanical peculiarities and

characteristics of the Celtic province are chiefly, almost entirely, marked by the inhabitants of the higher zones. This is strikingly exemplified by the mollusca, especially by those genera of them which are represented in the Littoral and Laminarian zones only; as *Patella, Purpura, Littorina, Otina, Lacuna, Scrobicularia,* and *Donax.* Similar instances might be adduced from among fishes, articulate animals and radiata. The comparatively few genera which have their species entirely confined to deeper zones within this area, extend in other regions to the shallower belts.

The inhabitants of the median or coralline zone around the British shores are numerous and various, but scarcely so peculiar as those of the preceding belts. Yet the general assemblage presents an unmistakable aspect of its own. Shell-fish, especially carnivorous mollusks, the whelk tribe above all, abound throughout it, varying numerically according to the nature of the sea-bed and the amount and kind of prey furnished by their hunting grounds. Bivalves of considerable beauty, especially clams and scallops, are found buried in numbers in its gravels and muddy sands, and Sertularian zoophytes throng so as to form miniature gardens, and around their graceful branches crawl and hang diversified kinds of worms and nudibrancheous sea-snails, not unfrequently of considerable beauty. The spider crabs are here plentiful, with many peculiar crustaceans. And, as a

THE EUROPEAN SEAS. 101

natural consequence of this accumulation of good food, fishes abound, and many of our deep sea and white fisheries owe their value to the zoological features of the coralline zone.

The abyssal regions of the Celtic seas are scarcely included within their more characteristic portions. The depths of ocean that bound the shallower soundings along the western side of Ireland and Scotland would, doubtless, if carefully explored, reward the naturalist richly for his labour, if not with new or extraordinary forms, at least with a knowledge of facts, desired, but not yet obtained. Some indications of the conditions of animal life in the Atlantic depths near our shores were obtained by Captain Vidal during his deep-sea survey, and, such as they were, held out good prospect.

It would seem that the tribes of annelides of the genus *Ditrupa*, tooth-like shells, very similar to *Dentalium* in their shapes, are especially abundant. The approach to land at the entrance of the Channel has long been inferred by mariners, from the presence of the shells, called Hake's teeth (*Ditrupa Gadus*), among the soundings on the lead obtained from deep water. It is probable that there is little difference between the fauna of the great depths hereabouts and that of the abysses of the Mediterranean, and we may hope, by their exploration, to track the course from north to south of certain species (such as *Limea Sarsii* among the mollusca) that have not as yet been noticed in the interval. But deep-sea dredging is at all times a difficult operation, and amid the roll of the Atlantic demands a good boat, plenty of zeal and leisure, unusually fine weather, and a strong stomach for its successful execution.

The aspect of the Celtic fauna is peculiar and modest. The shapes of its constituents of different tribes for the most part are but slightly diversified by eccentricities, and their hues seldom glaring or even vivid. The smaller kinds of sponges are not unfrequently brilliantly dyed, especially a few species of vermilion or golden yellow hue, but the more conspicuous kinds are tawny or brownish. The sea-anemones are elegantly variegated with rich colours, but the majority of zoophytes are not strikingly tinted. The Starfishes, as a group, are most remarkable among the invertebrata for gorgeous painting, but our other echinoderms are sombre when compared with their relatives from warmer seas. The sea-jellies are occasionally tinged with delicate hues, and some of the smaller kinds even showily ornamented; but those which make most figure in our waters are not conspicuous on account of colour, however elegant in their contours. Taken as a class, our mollusks are like the men and women of the lands around their habitations, very neatly but not gorgeously attired. The patterns of their shells, though often pretty, are not gaudy or attractive, except in rare instances. The same may be said, with slightly lesser truth, for our marine articulata. On close inspection, however, the elegance of device on the carapaces of many of our crustaceans is exceedingly admirable.

The fishes of the Celtic seas are not remarkable for brilliancy of painting. Their hues are quaker-like, though sufficiently lustrous for sober tinting. The Cod and Flounder tribes are among the most characteristic, and such of the more common fishes as belong to families of which we have but few representatives, are in most instances clothed in sober grey and silver. Beauty of no mean charms may, however, be displayed by these modest vestments; witness the mackerel and the herring.

Among the Celtic Wrasses are several exceptions to this rule; gorgeously decorated fishes. But these belong to a family more characteristic of seas to the south; for though there are a dozen or so species of *Labridæ*, haunting the mid-western coasts of Europe, more than thrice that number are indigenous to the Mediterranean. A like deficiency in the numbers of *Sparidæ*, *Triglidæ* (the Gurnet tribe), and *Scombridæ* (the Mackerel tribe), seriously affects the showiness of aspect of our piscine fauna, when compared with that inhabiting the Mediterranean. The Sharks and Rays too are comparatively deficient, although a few species are over-sufficiently abundant. The sea-eels are also few, although in the common Conger and the larger Sand-eel (*Ammodytes lancea*) we have two very characteristic Celtic species. The sea-perches are few, and the dolphins absent.

Among the Blenny family we have in this region the southern limits of the Gunnels, the viviparous Blennies, and the Cat-fishes, and the whole tribe of Cottoids attains its equatorial limit, so far as the northern hemisphere is concerned.

Within the British section of this province we find distinct indications of a transition, as it were, from a northern to a southern type. Several characteristic boreal forms find their southern limit within the northern half of the British area, and there some of the most striking and abundant kinds are chiefly developed in numbers, such as the cat-fish (*Anarhicas lupus*), the seythe (*Merlangus carbonarius*), the ling (*Lota molva*), the cod (*Gadus morrhua*), the lump-sucker (*Cyclopterus lumpus*), and even the herring (*Clupea herengus*). On the other hand, along the southern shores of England we find fishes becoming frequent that are distinctly of a southern type, such as the red mullet (*Mullus barbatus*), the sea-bream (species of *Pagellus*), and far more plentifully, the John Dory (*Zeus aper*), and the pilchard (*Clupea pilchardus*).

But although the Celtic province cannot boast overmuch of the beauty of its ichthyological subjects, when as yet unboiled and swimming free in the briny waters, it can challenge the world to match, if it can, its favourite and abundant fishes when they have undergone the gastronomic ordeal.

The turbot, cod, whiting, herring, whitebait, mackerel and sole may fairly be summoned to bear testimony to this merit; and though the salmon belongs rather to the history of the land than of the sea, it may, since it is anadromous, be called as an additional and powerful witness. Besides all these too, there are not a few of our less known native fishes in the British seas that deserve the commendations of the judicious epicure.

CHAPTER V.

THE LUSITANIAN PROVINCE.

SPAIN and Portugal, of all European kingdoms, have served most scantily the cause of science, and have contributed but a very small quota to the army of naturalists. Indeed, until within the last few years our knowledge of their vegetation, a subject usually in advance of other branches of local natural history, was fragmentary and imperfect, nor are we indebted now for the most that we know to Iberian botanists, few of whom have laboured assiduously among the treasures of their native land. By English, French, Swiss and German explorers has the rich flora of the Peninsula been sifted. Much yet remains to be done before the terrestrial zoology of this region shall have been satisfactorily examined. If this be the state of science upon the land, we can hardly hope for better things at sea; and, indeed, there is no province of the European seas about which we know so little in detail as the oceanic margins of Spain and Portugal. Were it not manifest that the natural history region of which they form a part, embraces, ere it reach its southern boundaries, the seas around Madeira and the Canary Islands, where able naturalists have laboured diligently and successfully, our account of the fauna of the Lusitanian region would be in a great measure hypothetical.

The most important and extensive contribution to our knowledge of the invertebrate animals of the Atlantic coasts of Spain and Portugal which has as yet been made public, is the account of the dredging researches of Mr. Mac Andrew, communicated by that indefatigable friend of submarine research, to the natural history section of the British Association, during the meeting at Edinburgh in August, 1850. In this document a record is presented of all the species of mollusca, their precise depth, locality, and nature of the ground upon which they were taken, with notes of their relative frequency and abundance, and notices of the animals of other tribes found along with them. The stations examined, which are especially connected with the region under review, were the Bay of Vigo in Gallicia (the investigator had previously explored part of the coast of Asturias), Lisbon, and Cascaes, south of the rock of Lisbon; the neighbourhood of Faro in Algarve; various points between the mouth of the Guadalquiver and Cape Trafalgar, and the Straits of Gibraltar.

The general results of these researches may be stated as follows : on the north coast of Spain bordering the Bay of Biscay, we find littoral species of mollusks of decidedly Mediterranean types, and which do not range to the Celtic seas. A peculiar *Littorina*, *Chiton cajetanus*, *Pleurotoma Maravignæ*, and *Pollicipes cornucopia*, may serve as striking examples. Vigo Bay is a great arm or lough of the sea running inland in a mountainous country. It is 16 or 18 miles in length, and as deep as 25 fathoms in the mid-channel, with a muddy bottom. Its most striking zoological feature is the significant circumstance discovered by Mr. Mac Andrew, that instead of its fauna being characteristically of a Mediterranean, or rather Lusitanian character, as might be expected by its position, and by the nature of the marine fauna of the Spanish coast to the north and to the south of it, we find the assemblage of animals and plants inhabiting this *fiord*, to use the Norwegian term, mainly of a Celtic or British character. Its littoral or coast-line animals are especially of British types. Out of 200 species of testacea taken there, only 25 are forms which do not occur in the British seas. Some of these are, however, remarkable, and serve strikingly to indicate the difference between the Celtic and Lusitanian areas, such as *Tellina serrata*, two species of the beautiful genus *Solarium*, *Trochus Laugieri*, *Ringicula auriculata*, and two species of *Triton*, including the great *Triton variegatum*, or Trumpet whelk. About 28, on the other hand, are species which do not range to the Mediterranean. Some of them are characteristically northern as *Patella pellucida, Velutina lævigata, Trochus tumidus* and *cinerarius, Lacuna puteolus, Littorina littoreus* and *rudis, Purpura lapillus, Mactra truncata, Tapes pullastra* and *Pecten tigerinus*. Now it is very important to note that the majority of these are characteristic, and mostly gregarious, species of the Littoral and Laminarian zones; species, moreover, which could only be transmitted along coasts presenting a line of rock or hard ground; and that they are univalves, which, as a general rule, are less widely-diffused shells than bivalves. Mr. Mac Andrew expresses his conviction that "the marine fauna of Vigo, so far as the mollusca are concerned, is more nearly related to that of the British Isles than to that of the region in which it is situated." Among its more remarkable productions is a large reversed *Fusus*, which, though differing in some of its features from the fossil *Fusus contrarius*, nevertheless so closely resembles some varieties of that curious shell, that it is hard to believe it to be other than the same species slightly modified. The importance of the existence of a British colony, so to term it, of littoral shell-fish on the deep bays of Gallicia, depends on the geological bearing of the fact. It is certainly a most striking circumstance that we should find these creatures living on the coast of Spain, only on its most extreme western region, and in juxtaposition with the sub-alpine flora of the Asturian type, which is partially present also on the western coast of Ireland. Now two years before Mr. Mac Andrew's discovery I maintained the theory that during the epoch preceding the present—during that epoch to which the terms *glacial* and *pleistocene* have been applied, and most probably at the early stage of that epoch—there was an extension of the land of Europe westward, as far as or beyond the Azores, that the land so extended was continuous with or more likely contiguous to, the land of Ireland, and that over this extended land migrated an Asturian flora, whose fragments remain on the mountains of the west of Ireland, and are represented there by the peculiar Saxifrages, Heaths, strawberry-tree, and some other plants (the number has increased since I wrote) not found elsewhere in the British islands. I will quote from the memoirs referred to.* "The remarkable point concerning these (Irish) plants is that they are all species which at present are forms either peculiar to, or abundant in, the great peninsula of Spain and Portugal, and especially in Asturias. No existing distribution of marine currents will account for their presence, and even if there were plausible grounds for attributing it to the great current known as Rennel's, which sweeps the northern coasts of Spain, and strikes in its aftercourse against the western shores of Britain and Ireland, the plants in question, instead of being

---

* On the Geological relations of the existing Fauna and Flora of the British Isles, in " Memoirs of the Geological Survey of Great Britain," vol. i.

where they are, should be present in the southern districts of the countries bounding the English Channel—in the region of the Devonian flora, *where they are not*. Nor can we suppose that they have been conveyed as seeds through the air; for besides the important fact that they are all members of families having seeds not well adapted for such diffusion, and that the species of Compositæ, and other plants with winged seeds associated with them in Spain, are not present with them in Ireland; it would be very extraordinary if the winds which had conveyed them so far, had never, through, probably, a long series of centuries, conveyed them still farther, and diffused them in a country where there are abundance of situations well adapted for their habitation.

"The hypothesis, then, which I offer to account for this remarkable flora is this,—that at an ancient period, an epoch anterior to that of any of the floras we have already considered, there was a geological union or close approximation of the west of Ireland with the north of Spain; that the flora of the intermediate land was a continuation of the flora of the peninsula; that the northernmost bound of that flora was probably in the line of the western region of Ireland; that the destruction of the intermediate land had taken place before the Glacial period; and, that, during the last-named period climatal changes destroyed the mass of this southern flora remaining in Ireland, the survivers being such species as were most hardy, saxifrages, heaths, such plants as *Arabis ciliata* and *Pinguicula grandiflora*, which are now the only relics of the most ancient of our island floras.

"This, I admit, is a startling proposition, and demands great geological operations to bring about the required phenomena. With such a gulf as now intervenes between Ireland and Asturias, it may seem fanciful and daring to suppose their union within the epoch of the existence of the plants now living in both countries. What then are the geological probabilities of the question?

"During the epoch of the deposition of the miocene tertiaries there was sea—probably shallow—inhabited by an assemblage, almost uniform, of marine animals throughout the Mediterranean region (tertiaries of Cerigo, Candia, Malta, Corsica, Malaga, Algiers), across the south of France (Montpellier, Bordeaux), along the west of the peninsula (Lisbon, &c.), and in the Azores (St. Mary's). I speak to the uniform zoological character of this sea from personal examination of its fossils.

"During the miocene epoch, then, we can suppose no union of Asturias and Ireland. But at the close of the miocene epoch great geological operations took place: witness the miocene marine beds discovered by Lieutenant Spratt and myself, at elevations from 2,000 to 6,000 feet in the Lycian Taurus. The whole of the bed of this great miocene sea appears to have been in the central Mediterranean and west of Europe, pretty uniformly elevated. This then could, with every probability, have been the epoch of the connection or approximation of Ireland and Spain. My own belief is, that a great post-miocene land, bearing the peculiar flora and fauna of the type now known as Mediterranean, extended far into the Atlantic, past the Azores, and that, in all probability, the great semicircular belt of gulf-weed ranging between the fifteenth and forty-fifth degrees of north latitude, and constant in its place, marks the position of the coast-line of that ancient land, and had its parentage on its solid bounds. Over this land that flora of which we have now a few fragments in the west of Ireland, might with facility have migrated. This would give us a new antedate, and enables us to declare our entire existing terrestrial flora and fauna as post-miocene."

This argument I further supported from the evidence of the fossils found in the drift (the upheaved bed of the glacial sea) of the south of Ireland. "The abundance of *Purpura lapillus,* and the presence of *Littorina littorea,* may be mentioned as especially characteristic of the shelly gravels which in Wexford have been found by Captain (now Colonel) James to contain numerous specimens of the reversed variety of the *Fusus antiquus,* known under the name of *Fusus contrarius,* and common in the red crag. At present the reversed form is as rare among specimens of that *Fusus,* as the dextral form was anciently. It is difficult to conjecture a sufficient cause for the prevalence of the monstrous over the normal form during two geological epochs. The discovery, by Colonel James, of *Turritella incrassata* (a crag fossil) and a Spanish species, of a southern form of *Fusus*, and of a mitra allied to [probably identical with] a Spanish species in these southern Irish beds, associated with the usual glacial species, is an important fact, suggesting the probability of a communication southwards of the glacial sea, with a sea inhabited by a fauna more southern in character than that now existing in the neighbourhood of the region where those relics were found."

I still stand by these opinions, after a full consideration of the many objections, some weighty and worthy of consideration, some frivolous and personal, which have been offered to my theory, to this part of it in particular, both at home and abroad. These shall be answered fully in due time; at present I prefer occupying myself in fresh research to wasting time in retrospective controversy. To that theory, I, however, recall attention here, since the Gallician discoveries of my indefatigable friend Mr. Mac Andrew, go most importantly to support my views. Let the peculiar distribution and presence of the littoral mollusca, before mentioned, on the coast of Gallicia, be explained (always bearing in mind my premisses respecting the unity of species) by any other view than that I advanced without the aid of these fresh and important facts, if they can.

At Cascaes Bay, south of the Rock of Lisbon, Mr. Mac Andrew dredged one of the most interesting and peculiar members of the Lusitanian fauna, viz. the *Cymba olla*, the only volute shell found in the European seas, and one of the largest of our mollusks. It was taken alive on a bottom of hard sand at a depth of from 15 to 20 fathoms. It ranges to low-water mark, and occurs abundantly in the south of Portugal. This beautiful mollusk is of a strikingly tropical aspect; it does little more than just enter the Mediterranean (I have picked up a dead young specimen as far as the shore of Algiers), and is abundant on the north-western coast of Africa, to which region (the Senegal province) it probably most strictly appertains. Out of a large list of shells obtained at Faro the following may be selected as strikingly marking the character of the region:—*Petricola lithophaga, Panopœa Aldrovandi, Psammobia* (*rugosa*-like species), *Ervilia castanea, Mactra helvacea, Cardium rusticum, Mytilus minimus, Solecurtus strigillatus, Bornia corbuloides, Natica intricata* and *Guilleminii, Phasianella intermedia, Trochus Laugieri* and *canaliculatus, Turbo rugosus, Cerithium vulgatum, Murex corallinus, trunculus, Brandaris,* and *Edwardsii, Triton variegatum, corrugatum* and *cutaceum, Purpura hæmastoma, Cassis saburon? Columbella rustica,* a large yellow *Mitra,* and *Conus Mediterraneus.* Out of 99 species enumerated, 59 or 60 are British species, but all, except *Trochus lineatus* (doubtfully determined) and *Trochus umbilicatus*, such as range into the Mediterranean, and many of them are found only on the southern shores of Britain. These were taken in shallow water, within the littoral and laminarian zones. The record of a dredge off Cape St. Maria, in the neighbourhood of the same locality, shows the character of the molluscan fauna of the coralline zone, having been worked in between 15 and 30 fathoms, on a bottom of coarse sand, and, in places, of mud. Out of 83 species enumerated, 59 are British, but the same remark applies to them as to those just mentioned from Faro. Among the remainder a few of the most striking may be specified: *Tellina distorta* and *costæ*, *Cytherea venetiana*, *Cardita trapezia*, *Lucina digitalis* and *divaricata*, *Mytilus afer*, *Leda emarginata*, *Pecten polymorphus*, *Natica sagra?* *Turritella sulcata*, *Buccinum modestum*, and *Ringuicula auriculata*. In the record of a dredge in 30 fathoms, eight miles or more from shore between Cadiz and Cape Trafalgar, we find a *Vermetus* and *Fusus corneus* (i. e. *lignarius*) taken, and the large red *Oculina* coral. Out of 265 species of estacea obtained in the Bay of Gibraltar, 135 are British species. In this list we find species of the genera *Solemya*, *Mesodesma*, *Cardita*, *Bornia* (as distinguished from *Kellia*), *Siphonaria*, *Vermetus*, *Solarium*, *Turbo*, *Cancellaria*, *Ranella*, *Triton*, *Cassis*, *Columbella*, *Rin-*

guicula, *Mitra*, *Cymba*, *Marginella* (as distinguished from *Erato*), and *Conus*, none of which are present in the Celtic fauna.

We find also the lost traces of some northern forms, as *Venus striatula*, *Pecten maximus*, *Ostrea edulis*, *Acmœa virginea*, and *Littorina neritoides*. With the exception of the last-named species the peculiar littoral assemblage of Testacea which holds its place from Nordland to Finisterre, and reappears, as we have seen, for a space in Gallicia, has entirely disappeared.

The Echinoderms of the Lusitanian seas are kinds common to the Mediterranean, and not Celtic species. The *Echinus esculentus verus* is the characteristic sea-urchin.

Probably in the present state of our knowledge the most marked distinctions between the Lusitanian and the Celtic regions are to be founded on the Testacea. The presence of members of the series of genera I have mentioned above, is especially a most unmistakable distinction, besides the numbers of species which do not range northwards of the Peninsula. The general assemblage of species, especially those which inhabit the littoral and laminarian zones, presents a much more gay and gaudy painting than in more northern seas. From the Mediterranean region, on the other hand, a number of peculiar Testacea, absent there, afford a distinction. Such are the *Chiton fulvus*, a large and singular species, which, contrary to the usual habits of its congeners, creeps on a sandy sea-bed, and which ranges from Gijon to the extreme south; *Cymba olla*, the great volute already noticed; *Lithodomus caudigerus*, a curious boring mussel, which takes the place here of the Date-shell (*Lithodomus lithophagus*) in the Mediterranean, and which has a range equal to that of the *Chiton fulvus; Psammobia rugosa, Siphonaria concinna, Turritella sulcata*, and *Mytilus afer*. We may also cite *Trochus umbilicatus*, a species characteristic of the oceanic shores of Europe from the north-west of Scotland southwards. The absence of the common Mediterranean *Chiton siculus*, on the other hand, a species which, if present, was not likely to have escaped the researches of the indefatigable explorer to whom I am indebted for so much of this information, is a significant negative fact.

There is evidently a fine field for original research unexplored in this portion of Europe. The sea-weeds of the shores of Portugal have recently been collected and distributed, but a large section of their zoology is almost or quite unknown. A good account, or, indeed, any pretty full catalogue of the fishes of the Portuguese coast is very much to be desired. We hope that before long some of our naturalists will direct their attention to this interesting and promising region.

To get a notion of the ichthyology of the Lusitanian province, we are obliged to travel out of European bounds, and have recourse to the excellent researches of the Rev. R. T. Lowe in Madeira. There is no catalogue of Portuguese or Oceanic Spanish fishes published, so far as I am aware. In 1837, Mr. Lowe communicated his Synopsis of the Fishes of Madeira to the Zoological Society of London. Of spiny-rayed osseous fishes he had observed 73 species in this locality. Of these 25 were peculiar to Madeira; of the remainder 26 were common to Madeira and the Mediterranean, 8 to Madeira, the Mediterranean, and the British seas, and 4 to Madeira and the British seas only. The species peculiar to Madeira were mostly sea-perches, sciænidæ, mackarels, and wrasses. Of soft-rayed osseous fishes, 20 Madeiran species were observed. Of these 7 were peculiar to Madeira, 8 common to Madeira and the Mediterranean, and 5 extending their range to Britain. Of the pipe-fishes, 2, one being peculiar, one Madeiran. Of Gymnodonts, the *Diodon reticulatus* and the *Tetrodon marmoratus* occur. Of file-fishes, one is found, a Mediterranean form. One member of the sturgeon tribe is present. Out of 12 sharks and 5 rays, 8 are common to the Celtic and Mediterranean provinces, and 6 range only to the Mediterranean in Europe. The total number of sea-fishes enumerated is 116; but this list has, we believe, through Mr. Lowe's subsequent indefatigable researches, been considerably increased.

The general aspect of the Lusitanian ichthyology may, perhaps, be fairly judged of from these data.

The number of kinds of fish must be regarded, compared with the Celtic number, as proportionately large, when we consider the limits and peculiarities of the district submitted to exploration. The spiny-finned division of the osseous fishes is especially well represented. Their proportion, as compared with the soft-rayed division, is greater than in either the British or Mediterranean seas. Mr. Lowe remarks, that, "instead of occupying a place, considered ichthyologically, corresponding with its latitude, Madeira seems to be intermediate between Great Britain and the Mediterranean." This would accord with our view of its forming a portion of the Lusitanian province. The number of fishes of tropical forms is smaller than we might expect from the position of the island.

The remarks made by Mr. Lowe on the facies of the Madeiran fish-fauna are so interesting and to the purpose that I think it well to extract them entire, especially as they may serve to interest and inform many of our invalid countrymen who may visit hereafter, in their search after health, the beautiful island whose marine productions have been so admirably investigated by this distinguished and accurate naturalist.

"The list of fishes," he remarks, "fails to convey a faithful picture of the general character and aspect of Madeiran ichthyology. It does not sufficiently express the decided predominance of the Sparidal, Scombridal, and Percidal forms above all others. This arises from the profusion in which the individuals of certain species in these families occur; while the species which compose the other families are in general poorer considerably in this respect. The commonest edible fishes of the island are found in the three families just named, as well as the more gregarious and prolific species.

"Thus the European visitor on entering the markets, or examining the boats, is struck at once with the almost total absence of the flat-fishes, salmon and cod-fish tribes, which more especially characterize our stalls in England, and with the unwonted form of the *Sargus, Pagrus, Box, Oblada, Smaris, Thynnus, Prometheus, Lichia,* &c., or with the brilliant hues of the *Serranus, Beryx, Acarus,* &c., or the grotesque deformed *Scorpœna* and *Sebastes.*

"This impression will be somewhat different at different seasons. The spring is characterized by the common appearance of the splendid-coloured *Beryx* in the streets; attracting notice no less by its form and hues of silver, scarlet, rose, and purple, than by the extraordinary size and opaline or rather brassy lustre of its enormous eyes. With this, or even earlier, appears abundantly the common herring of Madeira (*Clupea Madeiriensis*); and as the season advances the mackarel (*Scomber scombrus*); the scarlet Peixe Cao, or dog-fish of Madeira (*Crenilabrus caninus*); Carneiro, or mutton-fish (*Scorpœna scrofa*), and Requime (*Sebastes Kuhlii*);

the pike-like Bicuda, or spet of the Mediterranean (*Sphyræna vulgaris*); the Sargo (*Sargus Rondeletii*), with teeth resembling the human; the elegantly golden-striped but worthless Salema (*Box Salpa*), and the plain-coloured Dobrada (*Oblada melanura*). The herring and the Alfonsin (*Beryx splendens*) attain the climax of their season about March or April; the mackarel in May and June; but the whole, except the herring, continue throughout most part of the summer and autumn. In May the magnificent *Lampris lauta*, the beauty of which in the water excites the admiration even of the fishermen, begins to make its occasional appearance in the market; and what is of far more importance in an economic point of view, the Tunny fishery begins. This last is at its greatest height in June or July; and to it succeeds the capture of the Gaiado (*Thynnus pelamys*), which is pursued with such success, that I have sometimes watched a single boat, furnished with scarce half a dozen rods, pulling them in at the rate of three or four a minute. With the Gaiado appears in almost equal plenty, the Coelho, or rabbit-fish (*Prometheus atlanticus*), and these continue till the close of the summer by the equinoctial rains of October. The winter months of January and February are chiefly characterized by the presence, close along the shores, of the little Guelro (*Atherina presbyter*), or sand-smelt of Madeira, of the common Madeiran herring, and Sardinha (*Clupea sardina ?*); the two last being captured, principally,. after violent gales and storms, when the swollen rivers or torrents carry much mud into the sea.

"The following species occur in great profusion, more or less, throughout the year, but still most plentifully in the spring and summer; viz. Garoupa (*Serranus cabrilla*); Cherme (*Polyprion cernium*); Pargo (*Pagrus vulgaris*); Boza (*Box vulgaris*); Bocairao (*Smaris Royeri*); Ranhosa, or Tronbeta (*Lichia glaucos*); Chicarro, or Madeiran horse-mackarel (*Caranx Cuvieri*); Bodiao (*Scarus mutabilis*); and Abrotea (*Phycis Mediterraneus*). The well-known John Dory, or Peixe Gallo (*Zeus Faber*), and the delicate red mullet or Salmoneta (*Mullus surmuletus*), are also taken at all seasons, but more sparingly. The grey mullet, or Tainha, is captured very plentifully throughout the year, but most abundantly, perhaps, in June."*

How far the Lusitanian region may be said to extend westward into the ocean, is not yet determined. If my speculations regarding the ancient condition of this area be correct, the whole sea as far as the Azores, and those islands themselves, distant though they be, full 500 miles, from the coasts of Portugal, should fall within its bounds. That the terrestrial flora of the Azores is most intimately related by almost every species to that of the Peninsula, and to the islands on the north-western part of the African continent, mainly to the former,

* "Zoological Transactions," vol. ii. p. 199.

we know from authentic records. Unfortunately our knowledge of the indigenous animals of the land, and the creatures which live in the seas and on the shores of the western islands, is not so complete,—indeed is, for all purposes of geographical comparison, singularly deficient. This, then, is a field, if properly treated, open to important discovery, and for an energetic naturalist, sufficiently versed in marine zoology to qualify him for the task, having time at his disposal and the means to meet the expenses which the nature of the investigations would demand, there can scarcely be a nearer, pleasanter, and more compact district for monographic study. The only creatures of the natural history of the Azorean seas that have attracted attention are the Medusæ, which appear to abound in their neighbourhood. These creatures would seem to accumulate here in vast numbers. Lieutenant Wilkes, of the United States Navy, the energetic and able conductor of the great American exploring expedition, ingeniously suggests relations between this gathering of the floating radiata and the habits and distribution of the sperm whale, an animal which is fished in the neighbourhood of the Azores.* He remarks that these islands lie in the course of the great north polar stream, and form an obstruction to its passage, arresting and accumulating the creatures which constitute the whale's food. The Medusæ, thus swept south-

* "Am. Ex. Exp. Narrative," vol. v. p. 482.

wards, seek strata of water of the temperature best suited for them. The waters of the polar current are superficial in this region. The whales feed near the surface, instead of diving down to seek their food, as they do in higher latitudes. Medusæ will be borne to lower latitudes in greater abundance at one season than at others, according to the variable extension and force of the polar current, and the whales will follow them, changing their haunts accordingly at different seasons. This may to a certain extent be true, but not wholly so ; for in the first place, it is not the whale of the Arctic seas, but the sperm whale, which is present here ; and in the second, the experience of sea-going naturalists is every day proving more and more that Medusæ, although free swimmers in the ocean, are as definitely limited in their geographical distribution as more fixed animals ; so that the Medusæ of the Azores are not likely to come from the north. Indeed this fact seems to have attracted the attention of sailors ; I recollect meeting with a paper in the "Nautical Magazine," in which it was proposed in some circumstances to find the ship's position by means of Medusæ.

The southern limits of the Lusitanian province are extra-European. I had always fancied that the line might be traced at or a little to the north of the Canaries, until Mr. Mac Andrew returned from his cruise to those islands in 1852. The zoological volume of the great work by Webb and Berthelot, on the Natural History of the Canaries, had left that impression, especially so far as the marine invertebrata were concerned. Thus, among the shells enumerated by Alcide D'Orbigny in that important publication, are several tropical species of *Conus*, and other Senegal forms. It would now seem, however, that the line of boundary must lie to the south of the Canaries. The greater portion of the fauna consist of Spanish and Mediterranean forms. Among 270 species of mollusks, all except a small proportion are of this category, and no fewer than 80 or more are British forms. Some remarkable forms of *Scalaria*, *Aclis*, and *Pleurotoma*, seem to characterize the province. Among 125 species of shells dredged off Madeira about 100 were Mediterranean, and of these 58 ranged to the British seas. Some curious features are presented by the productions of the African coast at Mogador. In the harbour there Laminariæ are as abundant as in our own seas, and on the fronds of these sea-weeds lives *Patella pellucida* as with us. Out of 98 species of shells dredged there, no fewer than 54 proved to be British species, and 90 out of the entire number were Mediterranean forms. The Echinoderms of the region around the Canaries are mostly European. We meet, however, at Madeira for the first time large and beautiful sea-urchins of the genus *Astropyga*.

CHAPTER VI.

MEDITERRANEAN PROVINCE.

IT has already been intimated (p. 16) that the Mediterranean is not entitled to take rank as an independent marine province in respect of any very definite assemblage of original forms which have seemingly been called into being there; yet the interest attaching to this area, viewed zoologically, is so varied as will ever require that it should receive separate and special notice in the Natural History of the European Seas. There are its well-defined limits, the richness of the assemblage of forms which it contains, the extent to which from early times these forms have been collected and described, the ready access we have to a large portion of its coast-line, together with the facilities which its tranquil waters offer for the investigations of the naturalist. Again, with reference to the past, the genealogy of a vast number of forms, or the relation of the present fauna to a former one, the directions and extent to which the migratory movements of large assemblages of marine animals have taken place there, the modifications which certain forms have experienced in the course of such changes, are all of them points which there receive such illustration as to make the Mediterranean basin and its contents more suggestive to the naturalist and the geologist than any other sea with which we are as yet acquainted. It is to such considerations as these that the present chapter will be mainly devoted.

It will doubtless be a difficult matter with some naturalists to divest themselves so entirely of old prepossessions as to regard the fauna of this great internal sea merely as a subordinate and derivative one; such, however, it essentially is, and if we have heretofore viewed it otherwise, it has been owing doubtless to the circumstance that it has been so long known. It was on this account that with the rise of the present school of natural history investigation it became a typical region— one to which reference was constantly made in all questions relating to the geographical distribution of European forms. Cuvier and Valenciennes, in their great work on Fishes, Deshayes, with respect to Molluscs, habitually speak of certain forms as ranging *from* the Mediterranean over a given region without, and so also with many others. This practice will probably be continued, nor will it be attended with any inconvenience, provided the expression does not mislead and induce the impression that the direction which an integral portion of the great Atlantic fauna has taken in its *diffusion,* was outwards *from* the Mediterranean, whereas it was the reverse.

The Mediterranean Sea, viewed physically, is a vast lateral extension of the Atlantic, and its fauna is a full development of the most typical portion of the Lusitanian zone or province of that great ocean. In connection with its dependencies, terminating in the brackish waters of the sea of Azof, it repeats, and on a vast scale, all the phenomena which have already been noticed touching the relation of the Baltic Sea to the Celtic zone. We should hardly have ventured, even now, to speak thus confidently of the relations of the Mediterranean fauna, but for the recent researches of our own countryman, Mr. Mac Andrew, whose dredgings from the Bay of Biscay to the Canaries, together with the reliance which may be placed on the determination of the numerous Testacea he met with, render his labours a most timely aid in such an inquiry as the present.

In most striking contrast to that scanty guidance which is offered by the indigenous naturalists of Spain and Portugal for the Lusitanian border of the Atlantic, is the host of able investigators by whom we are met so soon as we enter the narrow straits, and have passed within the great Mediterranean basin. Michaud, Risso, and more recently Jeffreys, conduct us along the shores of Languedoc, Provence, and Nice ; Olivi along those of the Adriatic. The littoral of Greece has been described by Deshayes and his brother naturalists, and the examination of the Ægean, from its shores to its greatest depths, by Ed. Forbes, produced not only a detailed local fauna, but showed that it admitted of definite bathymetrical distribution, and that marine animals have their zones of depth, just as plants have their regions of altitude. The bearing of these researches on the special investigations of the geologist have hardly yet been fully appreciated. The Mediterranean islands have not been passed over. Mac Andrew has reported on the Balearic group; Payraudeau on Corsica. Sicily and the coasts of southern Italy have been illustrated by the admirable works of Delle Chiaje, Poli, Cantraine, and Philippi. The Molluscous fauna of the Algerian seas, which may be taken as a type of the North African coasts, has been described by Deshayes.

The Eastern Mediterranean carries our retrospect to earlier labours. "This sea," says Ed. Forbes, "which furnished Aristotle with the subjects of so many of his admirable researches, is of no slight interest to the student of marine zoology. In the writings of the great founder of Natural History Science there are allusions to its shores which prove that he drew from them part of his information; it is consequently classic ground to the naturalist as well as to the scholar."

Though the character of the Mediterranean fauna be not distinctive, it is yet so far peculiar that the assemblage of forms which may be there met with will be found as a whole to be more typically Lusitanian than any from the Atlantic border of that zone, a result dependent on some of the physical features of this internal sea, which may be here noticed.

The portion of the Atlantic coast-line which may be taken as characteristically Lusitanian extends from the 30th to the 40th parallel of north latitude. The Mediterranean Sea, according to the estimate of Admiral Smythe, measures 2200 miles from west to east, with a breadth of 1200, giving 6800 miles of coast; but when the irregular outline of this sea is taken into account, with its great advancing peninsulas on the European side, its promontories, bays, and countless islands, its marginal line may be safely estimated at 13,000 miles, the whole falling between the latitudes which are Lusitanian as to fauna. A marine fauna, in all its elements, is immediately dependent on extent of coast; it is that assemblage of animal forms which is to be met with from the marginal line down to depths of seventy or eighty fathoms; so that, extent alone being considered, it will be readily seen what a wide field the Mediterranean expanse offers for the development of the fauna of a distinct Atlantic region.

The inequalities of the bed of the Mediterranean are great and abrupt; these, as well as the irregularity of its coast-line, favour the development of a wonderful profusion and variety of forms of life within narrow limits. It may assist the naturalist to state that these inequalities have been found to be connected with the configuration of the adjacent land, as was long since shown with reference to the island of Sardinia. On the coast of Nice, wherever the surface rises gently landwards, it will be found that the sea-bed is continued with a corresponding slope downwards, as off Ventimiglia; whereas in places where high lands come down to the coast, as from Monaco to Mentoni, the depths are great—immediately off Villafranca there is as much as 200 fathoms water.

In addition to these inequalities, which, it will be shown, have an important bearing on the character of the Mediterranean fauna, there is another physical feature which has to be noticed, inasmuch as it has been considered that it has exercised some influence on the distribution of that fauna. A submarine ridge extends from the south-western extremity of Sicily to the advancing headland of Tunis on the opposite coast of Africa, along which there is scarcely more than thirty fathoms water, so that the Mediterranean depression is made up of two great basins—an Eastern and a Western.

Commencing with those low forms of life which have so long occupied debatable ground between the animal and vegetable worlds, we find the "Sponge tribe" forming a very characteristic portion of the Mediterranean products. "Sponges," says Ed. Forbes, " are abundant in the Lycian seas. The more valued kinds are sought for about the gulf of Macri, along the Carian coast, and the opposite islands. Rhodes is the seat of one of the depôts of the sponges of commerce.

"The species which live immediately along the shore, near the water's edge, though often large, are worthless: these are of many colours; some, of the brightest scarlet or clear yellow, form a crust over the faces of submarine rocks; others are large and tubular, resembling *Holothuriæ* in form, and of a gamboge colour, which soon turns to dirty brown when taken out of the water; others again are lobed or palmate, studded with prickly points, and perforated at intervals with osculi. These grow to a considerable size, but, like the former, are useless, since their substance is full of siliceous spiculæ."

The larger kinds are not found deeper than thirty fathoms, and most of them within a third of that depth. A few small species live at very great depths, and one, a Grantia, was taken alive in the Gulf of Macri in 185 fathoms water.

The sponge of commerce (*Spongia communis*) is found attached to rocks at various depths, between three fathoms and thirty fathoms. When alive, it is of a dull bluish black above, and dirty white beneath. There are several qualities, possibly indicating as many distinct species. The best are taken from about the Cyclades.

The common sponge of the Eastern Mediterranean is said to occur in the Red Sea and Indian Ocean.

The *Tethyæ* are sponges rendered firm by containing numerous needles of flint throughout their substance; of these, one species, *T. lyncurium*, extends from our West British and Irish coast, along those of Europe, into the Mediterranean; many others of this tribe have a like distribution.

There are creatures to be met with in the waters of all seas, and mostly near the marginal line, which are so minute that the aid of the microscope is often required to show the existence of some of them, and which yet occur in such myriads on certain coasts that their remains become, literally, as countless as the sands. These animals are even now but little known, and the names under which they pass have been generally taken from the forms of their shelly structures: these are the *Foraminifera*, and the animals are the Rhizopods, closely allied to Sponges.

Such as occur in British seas, of which some sixty species have been noticed, are exceedingly minute, and our knowledge of the class has been mainly derived from Mediterranean species, where both the forms are more varied, and where some attain much larger dimensions. The waters of the Adriatic, in particular, swarm with these creatures, so that an ounce of sand from the coast at Rimini was found to contain no less than 6000 of these organisms.

M. A. D'Orbigny, the first naturalist who attempted to methodize the *Foraminifera*, recent and fossil, grouped them under sixty genera; of these forty-four are found in the Mediterranean, containing about 200 reputed species.

Such low forms of life, hardly coming within the range of man's vision, may seem to some to be not deserving of notice in such a rapid sketch as this is; but the genera of the *Foraminifera* have an ancestry in time, dating back to the earlier ages of the earth's history; and, minute though they be, their exuviæ have helped to build up vaster masses of solid sedimentary strata than any other animal forms. The conditions which these forms indicate, are, therefore, of the highest interest to the geologist.

Selecting those Mediterranean genera which are most prolific, we have—*Nodosaria*, containing fourteen reputed species, of which three are also British. *Dentalina* has eight, of which two are British; *Vaginulina* has eight, of which one is common to our fauna. Of *Textulariæ* there are fourteen; *Bulimina*, twelve; *Rotalina*, sixteen; *Cristellaria*, eight; *Nonionina*, nine; *Triloculina*, eight, of which one is British; *Quinqueloculina*, twenty; of which we have two.

So far as present observations go, the Rhizopods decrease rapidly both in numbers and in forms, as we proceed from south to north along the European shores of the Atlantic. The British species as yet identified with Mediterranean ones amount only to about six per cent. Of these are *Truncatulina lo-*

bata and *Quinqueloculina subrotunda,* which have seemingly a world-wide distribution.

M. Alcide D'Orbigny obtained forty-three species of Foraminifers from two small parcels of sand collected at Orotava and Teneriffe; of these, seven are well-known Mediterranean species; four are West Indian; the remaining thirty-two species are to be considered as peculiar to the Canaries; it is most probable, however, that many of these are common to the African coast. With the exception of one form, for which M. D'Orbigny created the genus Webbina, the aspect of the whole assemblage is European; all the genera are common in the Mediterranean, and many species, though considered to be distinct, are evidently very closely allied to well-known Mediterranean forms.

Of the habits of the living Rhizopods we as yet know but little; some are free, some live attached to marine plants; the great bulk of described and figured forms of Foraminifers have been found in coast-line sand. Such, however, cannot have been the condition of accumulation of those thick tertiary beds of Italy or France, so largely composed of these organisms. Many of the species which are found fossil in the Italian deposits occur, also, in the coast-sand of the Adriatic, and we must suppose that their light exuviæ are mostly carried outwards, and deposited in the tranquil depths of zones beyond those in which the animals themselves had lived. Ed. Forbes observed that *Foraminifera* were extremely abundant through a great part of the mud of his eighth region (which extends from 600 to 1380 feet in depth), and for the most part appeared to be species very distinct from those in the higher zones. Representatives of the genera *Nodosaria, Textularia, Rotalia, Operculina, Cristellaria, Biloculina, Quinqueloculina,* and *Globigerina,* were among the number. The difference here noticed between these deep-sea Foraminifers and the well-known existing species from the higher or marginal zone is curious; it would have been desirable that the comparison had also been made with the series from the Italian tertiary beds.

The Mediterranean Sponges, as seen through the clear waters of that sea, spreading over broad surfaces from the margin downwards, in all their varied colours and delicate structure, suggest, as they did to the older naturalists, that they are the mosses and lichens of the sea. This system of representation extends beyond these cryptogamic forms, and the analogies between flowering plants and some of those compound animals we have next to notice, forms the subject of one of Ed. Forbes's most original and happiest speculations.

The Sertularians are composite beings, built up by individuals, each of which concurs towards a common living structure; and the offices of these several individuals, and of their parts, correspond with those which produce the composite structure of a plant: each polyp answers to a leaf, and performs analogous offices towards the nutrition and increase of the common mass.

In plants, the reproductive organs—flowers and fruit—are converted leaves. The small bodies attached to the stems and branches of our common *Sertulariæ*, wholly unlike the other parts of these plant-like animals, are their reproductive organs—the vesicles containing the ova; it was shown by Ed. Forbes that each of these was a metamorphosed branch; and as this theory of plant-structure is constantly deriving support from those vegetable monstrosities where floral organs revert to leaves, so also does it happen that the Sertularian vesicles exhibit like cases of imperfect conversion. This close analogy gives to the ovarian pods a generic value, and defines the limits of the true Zoophytes to the exclusion of the *Bryozoa*.

The Zoophytes thus limited are to be met with throughout the Mediterranean in wonderful profusion and beauty.

Forms of *Sertularia*, *Campanularia*, and *Tubularia*, which are common on our British coasts, are found abundantly along the Atlantic shores of Europe, and thence many of them extend into the Mediterranean.

Of the *Anthozoa*, the *Tubipores*, so abundantly met with in the great Indian Ocean, and in the Red Sea, even at its northern extremity, are seemingly wanting in the Mediterranean: the *Alcyonidæ* are, on the other hand, very fully represented. *Lobu-*

laria palmata, of the Lusitanian zone and Mediterranean, is also found in the Red Sea. *Isis*, with its flexible horny axis and calcareous nodes, belongs chiefly to Eastern seas; one species, *I. elongata*, an Indian Ocean species, was, however, met with by Philippi in the Sicilian waters. The Red Coral (*Corallium rubrum*), though apparently not confined to the Mediterranean (for Ehrenberg met with it in the Red Sea), is in a high degree characteristic of it, from its great abundance; yet it is not equally distributed there. In the Ægean it occurs sparingly, and only as small specimens; it grows largest and most abundantly in the Sicilian seas, in the gulf of Genoa, about Corsica and the other western Mediterranean islands, as also on the Spanish coast.

The African Coral, though abundant and of large size, is neither so compact, nor is its colour as bright, as that of France or Italy. It will be thus seen that the Red Coral has a Western Mediterranean distribution. These polyp-structures are of slow growth; as much as ten years, it is said, are required ere Coral-ground which has been dredged over, is again productive. Perfect specimens, forming miniature trees, may sometimes be seen from Sicilian seas a foot and a half in height. If the Red Coral occurs beyond the Mediterranean in the Lusitanian Atlantic zone, it must do so much more sparingly; it was not met with by any of the naturalists who have explored the Canaries.

The genus *Antipathes* has several species in the Western Mediterranean; of these, *A. subpinnata* is also Lusitanian. In the Canaries it attains a foot and a half in height, being much beyond its Mediterranean growth. *Gorgoniæ*, too, are numerous, though it may be well doubted whether all the reputed species rest on sufficient characters; some, such as *G. placomus*, *ceratophyta*, and *coralloides*, are common to the Atlantic. *G. tuberculata* attains a great size in the Gulf of Genoa and off the Corsican coast, with a stem several inches in diameter. It is somewhat curious that Ed. Forbes did not meet with a single specimen of *Gorgonia* in all his Ægean researches.

There is a small Gorgonian Zoophyte, which is found attached to the so-called White Coral of the Neapolitan seas (*Oculina*), for which M. Philippi has proposed the name of *Berbyce*.

The *Gorgoniæ* seem to set the laws of geographical distribution at defiance: there are certain species which are said to be common to the Indian Ocean, the Mediterranean, and to both sides of the Atlantic; as many as six, however, which are found about the Canaries, are admitted by M. D'Orbigny as also Mediterranean.

The *Pennatulæ*, or Sea-pens, though local, are varied and numerous. *P. phosphorea* extends into our British seas, the other Mediterranean species are also Atlantic. Captain Spratt found two forms living in great abundance off the mouth of the Hermus. *P. setacea* is common to the Mediterranean and the Canaries.

MM. Quoy and Gaimard, who devoted a few days to the investigation of the marine fauna of the neighbourhood of Gibraltar, when starting on their great voyage in 1826, amongst many new and interesting objects, captured a magnificent specimen of a compound polyp, belonging to the genus *Veretillum*, consisting of a cylindrical body, more than a foot in length, yellow and orange, and studded with hundreds of white flower-like stars, each borne on a slender transparent stalk. These compound animals, so uninteresting and even repulsive when cast dead upon the beach, are, when living, amongst the most wonderful and beautiful of the strange things of the sea.

A remarkable Zoophyte, *Funicularia quadrangularis*, two feet and a half in length, taken by Mr. Mac Andrew off the west coast of Scotland, is also a Mediterranean species.

In none of their many zoological differences is the contrast between the Red Sea and the Mediterranean greater than in respect of their assemblages of Polyp animals. Of the *Actiniæ*, *A. mesembryanthemum* and *A. tapetum* are the only two species in common. This contrast is greatest as to reef-building corals; the Red Sea from end to end has literally been obstructed by them, but not only are they wanting in the Mediterranean, but are equally so over the whole of the European and African shores of the Atlantic. Bermuda has been built up by coral polyps; the islands on the old world side, such as the Azores and Canaries, are wholly without them. The stony corals of the European seas are few, insignificant, and solitary, but their distribution is very definite. Of the Turbinolids, *Sphenotrochus Andrewianus* of our western seas, and which I have found mid-channel as high as the meridian of the Isle of Wight, *Desmophyllum Stokesii*, and *Cyathina Smithii*, form our Celtic group. *Desmophylum cristagalli* makes its appearance in the northern Lusitanian zone, and *D. *stellaria, Cyathina *cyathus*, and *C. pseudoturbinolia* are Lusitanian and Mediterranean. *Cœnocyathus Corsicus* and *C. anthophyllitis* complete the Mediterranean Turbinolids. From a specimen I found in a Mount's Bay fishing-boat, I expect one of these last will prove to belong to our Channel fauna.

Of the Eupsammids, *Balanophyllia verrucaria* and *B. Italica, Dendrophyllia *ramea* and *D. cornigera* are Lusitanian Atlantic, as well as Mediterranean throughout, but the principal bulk of the forms of this order occur in the Southern Ocean, whence a few species range north on either side of the African continent, our Lusitanian forms being the remotest representatives.

Cladocera *cespitosa, C. stellaria, and *C. astrœaria* (a new species from the seas of Naples lately added by Sars), with *Astroides calycularis*, complete the Astreads of the European seas. Of the foregoing forms those marked with an asterisk are also met with in the Red Sea.

Cyathina pseudo-turbinolia and *Balanophyllia Italica* have had a long occupation of the Mediterranean region. In like manner our *Sphenotrochus Andrewianus* had early representatives in such fossil species as *S. milletianus* and *S. intermedius*.

Of the higher division of Medusæ, one—the Sea-blubber (*Aurelia aurita*)—ranges along the western shores of Europe, and throughout the whole of the Mediterranean; as many as half a dozen reputed species are perhaps referable to this our commonest form. *Pelagia*, an Atlantic genus, but of the Lusitanian zone, just reaches our south-west shores. They are abundantly Mediterranean, and in that sea are so phosphorescent at times, as to show like globes of fire beneath the waters. *Rhizostoma*, a rare form on our own British coasts, and *Chrysaora*, swarm in the Western Mediterranean.

Of the Medusæ we know but little as yet, either of their development, the functions of their several parts, or of their habits and distribution. Sea-going naturalists meet with them in greater numbers than any other forms of life; at times our internal seas, such as the Irish and English Channels, swarm with them; they float up into all our estuaries, and if we venture out into the open Atlantic, in advance of our western coasts, *Medusæ* may still be met with. Viewed in this way, certain forms are more pelagic than others, whilst at the same time we can see that there is a certain limitation, dependent on latitude.

Of the NAKED-EYED MEDUSÆ, *Oceania, Æquorea,* and *Geryonia,* have a great range. *Foveolia* and *Ægina* are Mediterranean forms, having southern relations. Our British forms of *Turris* and *Thaumantias,* which have a range north, have as yet been so seldom quoted from the Mediterranean as to show that the genus is sparingly represented there.

These animals (MEDUSÆ) swarm about the Straits of Gibraltar and the Western Mediterranean; those of the Adriatic, which have been well described by Professor Will, are still numerous, but they are scarce in the Eastern Mediterranean, where the absence of varied forms is balanced by the vast numbers of the common *Aurelia* which are there met with.

Velella and *Porpita.*—Medusa-like animals, with cartilaginous supports, belonging to the Lusitanian Atlantic zone, are Mediterranean; so also is *Stephanomia.* Numerous forms of these and allied animals were captured by the French naturalists Quoy and Gaimard, p. 141; but, strange to say, the beautiful Portuguese man of war, the *Physalia pelagica,* seldom passes from the open Atlantic. *Beroe,* with a considerable northern Atlantic range, is Mediterranean. Of this group, the most remarkable is the "girdle of Venus" (*Cestum Veneris*), from five to six feet long, and three inches broad, a long riband-like

Medusa, of translucent gelatine, fringed with a double row of cilia, which reflect lines of all delicate tints of light as it moves through the waters.

All these varied forms, as they are seen from a vessel's side, drifting along on calm, sunny days, suggest that they must be the sport of winds and currents, and be so wafted into all zones or latitudes. But such is not the case; and whether it is that, except on these calm and sunny days, they keep below, and so are not affected by the agents that bring southern forms of plants and animals into our British seas, still it is the case that the characteristic Lusitanian forms seldom reach us. It will be sufficient to compare the series of forty-five NAKED-EYED MEDUSÆ, described by Ed. Forbes as British, (a list which, for this purpose, might be curtailed,) with about a like number from the Mediterranean, to be satisfied how distinct are the *Medusæ* of these two regions. With respect to the *Arachnodermata* generally, the laws of geographical distribution are not only rigidly observed, but have also been somewhat closely drawn.

The Mediterranean *Bryozoa* require a somewhat detailed enumeration and notice. These are the forms, some of which, though they will appear under new names, have been long and familiarly known to our sea-side collectors as Corallines, but whose claim to take the higher rank of Molluscs has become universally admitted. Like some of the Polyps, these animals live associated in colonies, each forming a little cell; and it is in the definite shapes and modes of arrangement of these that characters are found by which the multitudinous assemblage of forms of *Bryozoa* can be systematically ordered.

Not only is this class of animals of great assistance in the determination of the relation of the Mediterranean to surrounding faunas, but the beauty and perfection in which their remains have been preserved from the earliest times, aid us materially in interpreting the evidence of change during the yet remoter past. M. A. D'Orbigny's primary division of the *Bryozoa* is into two great orders,—the Cellulinear, in which, to take outward characters, the cells are arranged end to end, or side to side; and the Centrifuginous, in which the cells spring from behind, or at the base of one another. These two orders are by no means equally represented among existing forms; some few of the latter occur in our European seas, but its representatives have, for the most part, passed away. In secondary and tertiary times, however, they swarmed in the seas now occupied by our Celtic and Lusitanian zones. At present the forms of this order have a wide distribution, and are known to reach high northern or southern latitudes. *Seriolaria unilateralis*, *S. convoluta*, and *S. lendigera* are Lusitanian and Mediterranean; the latter is also Celtic. *Crisea eburnea* has a great Atlantic range as low as the Canaries. *Cresidea cornuta* is

Atlantic and Mediterranean. *Myriozoum cuneatum*, common to the whole Mediterranean and so abundant in the Adriatic, is said by Ehrenberg to occur in the Red Sea. *Reptotubigera tubulifera, Entalophora proboscidea, Proboscina serpens*, and *Berenicea prominens* are Lusitanian and Mediterranean. The last two are found on the opposite sides of the Atlantic. One species of *Hornera* completes the Mediterranean series of Tubulinar and Foraminated *Bryozoa*.

Of the Cellulinear order, *Acamarchis neritina* includes the Mediterranean in its ubiquitous range.

Pherusa tubulosa, Reptoflustra impressa and *depressa* are Lusitanian and Atlantic. *R. membranacea* of our seas ranges north, but not south.

Reptolectrina dentata and *pilosa* extend from the Scandinavian region to the Canaries and into the Mediterranean. *Chelidonia cordieri* and *Aetea anguina* range from the Mediterranean to the Canaries; the latter is British.

Tubucellaria opuntoides is equally common in Sicilian seas as on the coast of Algeria; our common *Cellaria salicornia* reaches the Mediterranean.

Of the *Escharæ* of our seas, *E. foliacea* and *E. fascialis*, the latter reaches the Mediterranean. *E. cervicornis* is Lusitanian and Mediterranean; there are also forms of *Retepora* and *Semieschara*.

Hippothoa, with only a few species, and with a wide distribution, has one or two forms which have not been observed beyond the Mediterranean; the genus *Mollea* is also represented there.

Cellepora, of which *C. coccinea* is a British and northern representative, becomes amazingly abundant, numerically and specifically, in the Mediterranean seas. Of this genus, there are some eighteen species, most of which were discriminated by Delle Chiaje; some of them have been recognised in the Atlantic Lusitanian zone. *Celleporæ* are even more varied in the Red Sea, to which as many as twenty-five distinct species have been referred—these have, in some cases, Indian Ocean relations; the species of the two seas are essentially distinct.

Perina and its allied genera has several Mediterranean species.

The researches of Müller and Troschell, those of Ed. Forbes, both in the Celtic province and in the Ægean, together with the work of Grube on the distribution of the Adriatic and Mediterranean Echinoderms, have been the means of advancing our knowledge of this great order beyond that of some other portions of the fauna of this sea. From these sources the Adriatic Crinoids, Ophiurids and Asteriads may be estimated at about twenty-eight; the Holothuriads at seventeen. Sars, who has recently described the Neapolitan Echinoderms, finds, of the three first of these orders, as many as forty-five, of the Holothuriads thirteen, species.

Of the forty species noticed by Ed. Forbes as British, nine at least may be commonly met with throughout the whole extent of the Mediterranean, such as *Comatula rosacea,* **Ophiura lacertosa, Ophiocoma scolopendroides, Palmipes membranaceus,* **Asterina gibbosa (minuta),* **Asterias aurantiaca,* with its numerous varieties, *Echinus lividus, Spatangus purpureus, Echinocyamus pusillus.* To these, according to Grube, may be added that most singular in its aspect of all Echinoderms, *Astrophyton scutatum.*

These forms, in common with all the British species, have a considerable northern range, whilst in the contrary direction they extend through the Lusitanian zone, some even as far as the Canaries, where those species marked by an asterisk also occur. The European star-fishes, therefore, " do not seem so local in their distribution as the Mollusca and the higher classes of animals."

Taking only the common well-known and well-defined Mediterranean Echinoderms, it will be found that they are also Atlantic. *Echinus esculentus,* which does not reach our seas, is common along the west coasts of Spain and Portugal, as also on those of West Africa and the Canaries. Such, likewise, is the case with *Asterias tenuispina, Ophidiaster ophidianus, O. granifer,* and *Brissus ventricosus.* In some instances extreme zones of the Atlantic have forms in common, which, so far as we yet know, are wanting in the intermediate space; such is the great *Stellonia glacialis,* which is found in the

Mediterranean, and as far south as the Canaries, but which has not yet been recognised as a British species.

When the animals of different sea-zones are brought together and compared, it is constantly found that variation in size is a marked character with reference to species which are strictly identical; the British naturalist finds constant occasions for noting facts of this kind, when pursuing his researches away from his own immediate seas. With respect to Echinoderms, *Ophiura texturata* and *Echinus lividus,* from the south coasts of France, Spain, and the Mediterranean, exceed ours in size, and we have a still more remarkable instance if, as is supposed, the great *Echinus melo* be the same with our *E. sphæra.* On the other hand, *Spatangus purpureus* attains a much greater size on the coast of Norway than it does in the Mediterranean. Size and numerical abundance of any given form may be taken as the surest indication that it is at home there, or in its proper zone ; true northern forms degenerate and become scarce as they range south, just as southern ones do as they occur north. The directions in which such changes as these take place should be carefully noted, for these forms are not depauperized stragglers from their natural settlements, but rather the remnants, and indications of changed conditions, and are of the same value to the naturalist that the lingering communities of isolated races of man are to the ethnologist.

With us, the common "egg-urchin" affords the poor a somewhat stinted luxury; but in the Lusitanian area, and throughout the Mediterranean, its greater size, as also that of its allies, *Echinus melo* and *E. sardicus*, renders them, when "in egg," important articles of food. In Sicily they are in season about the full moon of March, there the *E. esculentus* is still called the "King of Urchins," whilst the larger Melon Urchin is popularly considered to be its mother; hence its name *Echinometra*, of the old naturalists. The size and abundance of these edible species is one of the striking peculiarities of the fish-markets of the Mediterranean seaboard.

Amongst the star-fishes of the Canaries, or of the south Lusitanian zone, are *Stellonia tenuispina, Ophidiaster ophidianus*, with its long, snake-like arms, *O. granifer*, and the large *Brissus ventricosus;* these all pass into the Mediterranean, as does also the *Cidaris imperialis*. This form occurs also in the Red Sea; but as a safeguard against any false inferences from such a fact, *Astropyga* has its northern limit about the Canaries, ranging thence down the African coast, and extending into the Red Sea in virtue of having a corresponding zone on the eastern side of the great African continent.

Echinus lividus is abundant in the Eastern Mediterranean, adhering to rocks a little below the water-mark. *E. esculentus* is found more sparingly and rather deeper, but the *Echinidæ* are not largely represented here. The *Echinus monilis*, an Atlantic species, but which had a Mediterranean settlement as far back as the oldest tertiary deposits, was found to be common at depths from fifteen to 200 fathoms. To these may be added the gregarious *Cidaris hystrix*, and *Spatangus purpureus*, *Echinocyamus pusillus*, the smallest and prettiest of our own urchins; a *Brissus* completes the *Echinidæ;* in all eight species, all Atlantic, and of which five extend into our seas.

Of the recognised *Asteriadæ*, these same seas contain our western "Spiny Crab-fish" (*Uraster glacialis*, Linn.) and the *Palmipes membranaceus*. The northern seas, observes Ed. Forbes, greatly exceed the Mediterranean in the number of species and abundance of individuals of this order. Out of the small number of the true star-fishes taken by him one-half occurred only as single specimens. So, also, with respect to the true urchins,—the edible species, so abundant in the central and Western Mediterranean, is individually scarce in the Ægean, as is also *Spatangus purpureus*.

The *Ophiuridæ* observed by Ed. Forbes in the Ægean were eleven; of these, four are Atlantic species; the rest, *O. texturata* and *albida*, *Amphiura neglecta*, and *Ophiothrix rosula*, are new, and were procured from great depths; one, the *Ophiura abyssicola*, having been taken alive from 200 fathoms.

Lastly, the *Holothuriadæ* are much more numerous in the Western Mediterranean than the Eastern.

They all live in shallow water, and attain a great size. *Cucumaria pentactes*, which reaches far north, was taken there, as was also the *Syrinx nudus*.

Some of the Echinoderms of the Mediterranean illustrate the changes of range in latitude which the same species exhibit when we compare present faunas with those of past times ; a large and beautiful urchin found in the Crag formation of our eastern counties is identical with the *Brissus Scillæ* of the Lusitanian regions.

Crabs, Lobsters, and Shrimps, belonging to the order CRUSTACEA, are rich, both numerically and in species, throughout the European seas, and they admit of a like geographical distribution to that which has been noticed with respect to other marine forms of life. M. Milne Edwards, who has devoted much time and study to these animals, was the first to present a sketch of this sort, and indicate those zoological provinces into which the European seas may be divided. Some species are peculiar to the Scandinavian region, others to the Celtic. The Mediterranean, again, contains species which are not to be met with in either ; so that, with respect to Crustacea, he considered that the European seas presented three distinct regions. The west coast of Africa has its peculiar Crustaceans, constituting a fourth region, and the Atlantic islands might, perhaps, form a fifth. This "Celtic region" of M. Milne Edwards is of much greater extent than that which Ed. Forbes designates by the same name, and includes a large portion of the Lusitanian zone.

Owing to the works of Leach, Desmarest, and others, we have long had a knowledge of the Crustacea of our own shores as well as of those of Brittany and the Mediterranean, to the exclusion of such as occurred on the south-western coasts of France, and the Atlantic border of Spain and Portugal. As M. Milne Edwards carries his Celtic region as low as Gibraltar, the assemblage from that region seemed to present an amount of distinctness from the Mediterranean which does not really exist.

The forms which may be considered Celtic in the restricted sense of the present work, are the Swimming Crab (*Polybius Henslowii*) of the west of France and England, *Hyas coarctatus*, *Athanas nitescens*, and *Pandalus annulicornis*. Other species, such as the Great Crab (*Cancer pagurus*), the common "Shore-Crab," *Carcinus mœnas*, and *Portunus puber*, have their numerical maximum within our region; its negative character, as a Zoological province, consisting in the absence or scarcity of *Catometopes*, *Anomoura*, and *Squilla*. In general terms, the Celtic Decapod Crustaceans of our coasts are to be met with in the Mediterranean; some of our common forms become scarce there, and *vice versâ*, indicating both the changes which take place across the Lusitanian zone, and the source or direction in which the Mediterranean has derived a large proportion of its Crustacea.

If we next separate the forms which may be considered, for the present, as characteristically Mediterranean, we find *Lupa hastata, Lissa Gualteri ?* * *Mithrax dichotoma, Herbstia nodosa, Amathia Rouxii, Acanthonyx lunulata,* several species of *Lambrus, Calappa granulata, Dorippe lanata, Hemola spinifrons (barbata), H. hispida,* several large forms of *Pagurus, Scyllarus latus,* and *Squilla mantis.*

Catometopes becomes numerous here, and certain southern genera make their appearance, such as *Ocypode ippeus,* abundant at Cape de Verde, *Gecarcinus,* &c.

The Decapod Crustaceans of the Mediterranean may be taken at ninety described species. Turning to the south Lusitanian region, as it is represented by the Canaries, we find there as many as forty species, of which a great proportion, with the exceptions to be noticed, occur in the Mediterranean. The south Celtic and the south Lusitanian Crustacea, together with a few from the west coasts of Africa, make up the assemblage of known Mediterranean forms. The direction whence this portion of the Mediterranean fauna has thus been derived is, therefore, evidently western. The known eastern Crustacea of this sea do not amount to one-half of those to be met with in the west; but, though less numerous as to species, their relations are still wholly western or Atlantic. Of the forty-three species of Decapods collected by the French naturalists on the coasts of Greece, one-third are British; and of the three species of Stomatopods we have two—*Squilla mantis* and *Desmarestii*.

The Mediterranean Crustacea are interesting in another point of view: the mineral composition of the external crusts of these animals favours their preservation; hence their remains are abundant in the old sea-beds of this area, and enable us to compare the relations of the present fauna to a former one, as readily as by the aid of fossil shells. Geological changes, and the influence they have exercised in the breaking-up of former zoological regions, or continuity of given forms, seem the simplest resource by which to explain the present apparent isolation of certain species.

The *Nephrops Norwegicus* has its numerical maximum in, and is a good characteristic Crustacean for, the Scandinavian region, but it occurs abundantly in Dublin Bay; it has not, however, according to Mr. W. Thompson, a general distribution—such as west and south, even throughout the Irish seas. We may feel sure, from its excellence as an edible species, that it has not been overlooked by fishermen, whilst its size, form, and proportions make it the most elegant Crustacean we have—a prize which no naturalist would overlook; yet, strange to say, it has not been recorded from the western coasts of France, nor do we meet with it till we reach the Mediterranean. It seems to be abundant in the Adriatic, in which sea it may be noticed, that several other outlying forms of northern types have also been met with.

Amongst the Mediterranean Crustaceans, there is found a species of *Mithrax*. The genus is characteristic of the western or American side of the Atlantic. When we consider the pelagic habits of some of the Crustacea, we might expect a much greater amount of agreement between the remoter portions of wide seas or oceans, so far as these animals are concerned, than could be looked for in the distribution of the Mollusca or even of the fishes : of all marine animals, certain Crustaceans are the most oceanic ; the central Atlantic regions of the floating weed-banks swarm with them. It will not, therefore, surprise us to find that the Crustacea of the Atlantic islands present an assemblage which departs a little from the Lusitanian character of the fauna of that group. Of forty-three species, more than half are also Mediterranean, and a few Celtic and Lusitanian, as *Inachus dorynchus;* but the distinctive character of the assemblage is derived from the southern forms. *Grapsus strigosus* and *Messor*, marginal crabs, are abundant, as is *Plagusia clavimana ;* these range down the African coast into the Southern and Indian Oceans, and occur in the Red Sea. The " Sea-spider," *Leptopodia sagittaria*, is common to the Canaries and the West Indian Islands.

The Tunicated Mollusks have not as yet been alluded to in the notices of the northern Atlantic provinces; those of our own Celtic coasts are somewhat numerous, but they are forms which, for the most part, do not occur very readily, whilst the more common ones, such as *Ascidia intestinalis* and *A. canina*, as they are usually seen attached to the roots of weeds thrown upon the beach, are but little attractive.

The geographical distribution of this class has not yet been worked out. Certain genera of simple Ascidians, such as *Molgula, Pelonœa, Boltonia*, and that very remarkable one, *Chelyosoma*, where the leathery skin is thickened into tortoise-like plates, seem to have a decided northern tendency. *Molgula* and *Pelonœa* are represented in our fauna, and our Ascidians as a whole have a northern distribution.

Some forms of this class have evidently a wide range; of our British species, *Amouroucium argus, Botryllus polycyclus, Ascidia mentula*, and *A. arachnoidea* are also Mediterranean.

There is a large Ascidian found in the Adriatic, in form somewhat like our common species, but which becomes a beautiful object from the effect of colour; in this—*A. papillosa*, the tough skin is thickly overset with disks of the brightest scarlet.

The genera *Eucœlium* and *Diazona* (a Medusa-like Tunicary) are characteristically Lusitanian and Mediterranean. The pelagic genus *Salpa* is represented here by more than a dozen species. Numerically these free swimmers are more abundant about the western than the eastern division.

There is a large compound Ascidian, easily to be mistaken for a Zoophyte, so abundant at times that the fishermen's nets become choked with it, and which, when thrown upon the beach, is amorphous and repulsive enough. When in the water it is seen to be a most remarkable form of aggregate life, consisting of a hollow cylinder closed at one end, and made up of hundreds of distinct animals set side by side: this is the *Pyrosoma*—the Firebody.

Though common, it has, perhaps, more frequently attracted attention at night than day, for, of all the numerous phosphorescent animals of Lusitanian seas, this is, perhaps, the most so. In particular states of the atmosphere, these animals light up the water by their intense fire-like glow. There are several species, but all belong to the warm regions of the Atlantic, and have their northern limits in the Lusitanian zone.

Many more products of the sea are eaten in southern regions than with us. *Ascidia microcosmus* is a favourite on the coasts of the gulf of Genoa, and *A. rustica* on those of Greece and the Adriatic.

The Mediterranean Pteropods belong mainly to the genera *Hyalœa*, *Cleodora*, and *Creseis*, forms wholly unknown to our own fauna except as waifs. Vast shoals of these animals frequent the deeper parts of that sea, leaving their remains strewed over its bed, between depths of 100 and 200 fathoms; they are short-lived creatures, and have their season, being met with near the surface during spring and winter, and were found by Ed. Forbes to have been most abundant from about three hours after noon till night-fall, "sparkling in the water like needles of glass."

These are the *winged* insects of the sea, reminding us, in their free circling movements and crepuscular habits, of the gnats and moths of the atmosphere; they shun the light, and if the sun is bright you may look in vain for them during the life-long day,—as days sometimes are at sea; a passing cloud, however, suffices to bring some *Cleodoræ* to the surface. It is only as day declines that their true time begins, and thence onwards the watches of the night may be kept by observing the contents of the towing-net, as the hours of a summer day may be by the floral dial. The *Cleodoræ* are the earliest risers; as the sun sets, *Hyalæa gibbosa* appears, darting about as if it had not a moment to spare; for its period is brief, lasting only for the Mediterranean twilight. Then it is that *Hyalæa trispinosa* and *Cleodora subula* come up. *Hyalæa tridentata*, though it does not venture out till dusk, retires early, whilst some species, such as *Cleodora pyramidata*, are to be met with only during the midnight hours and the darkest nights. This tribe, like a higher one, has its few irregular spirits, who manage to keep it up the whole night through. All, however, are back to their homes below before dawn surprises them.

In the descriptions of the other provinces, no notice has been taken of that large and highest order of Mollusca—the Cephalopods, or Cuttlefishes; they form, it is true, no insignificant proportion of our Celtic marine fauna, amounting to fourteen species, yet some of these can hardly be considered as more than the summer visitants of our seas, nor moreover are there any which are peculiar to our province. As an order, the Cephalopods increase in numbers and in representative forms as we proceed from cold to warmer regions, so that their history properly belongs to the Lusitanian zone of the European seas. These animals have been described in the general work of M. D'Orbigny, whilst the forms which occur in the Mediterranean are the subject of a monograph by M. Verany, of which the illustrations are the truest and most beautiful representations which have ever been given of these forms : from these sources, and by the notices of some few other naturalists, the number and distribution of the species of the order belonging to the European seas, may be easily determined; it may be fairly doubted, however, whether our knowledge now is not relatively far less complete than it is with reference to many of the lower orders of Mollusca.

The Cephalopods are migratory animals, wily and cautious, quick-sighted, rapid in their movements : many are pelagic, perhaps nocturnal also ; the little *Spirulæ* must swarm somewhere in Lusitanian latitudes, for their shells are brought by thousands to the coast of the Peninsula, yet they are never captured alive there.

Philippi met with fifteen species of Cephalopods in the seas of the Two Sicilies; of these, the large and ubiquitous "poulp" (*Octopus vulgaris*) occurs abundantly in the Eastern Mediterranean; but two other species (*O. velifer* and *O. catenulatus*) have not been taken, neither was *Argonauta argo*, though there is reason for supposing that it occurs there. On the other hand, *Eledone macropodius*, an abundant Greek species, and which was captured by Ed. Forbes at Cerigo, is not noticed by Philippi. The more open-sea researches of Ed. Forbes may perhaps account for some of the differences in the lists of these two naturalists, for in another place ("Travels in Lycia," vol. ii. p. 100), we find that *Octopus, Eledone, Sepia, Sepiola,* and *Loligo* occur in the Eastern Mediterranean; this is exactly the generic assemblage given by Philippi.

From an inspection of all the lists of various observers, we may fairly infer that the Cephalopods are scarcer in the Eastern Mediterranean than they are in the central portion.

Philippi is of opinion, that some of Verany's more western species (from the gulf of Genoa) may be found in Sicilian seas, though from their *scarcity* he had failed to meet with them. Apart from specific forms, the Cephalopods of the Eastern as compared with those of the Western Mediterranean, illustrate what happens with respect to marine animals generally ; where the distinct forms are few, the individuals are numerous, and where they are more varied, there the common forms are individually less abundant. About the shores of the Eastern Mediterranean the common *Sepia officinalis* is so numerous, that the " cuttle-bones " may be seen in places heaped by the waves into a ridge, which fringes the sea for miles. " As in ancient times," says Ed. Forbes, "these mollusks constitute now a valuable part of the food of the poor, by whom they are mostly used. One of the most striking spectacles at night on the shores of the Ægean, is to see the numerous torches glancing along the shores and reflected by the still and clear sea, borne by poor fishermen paddling as silently as possible over the rocky shallows in search of the cuttle-fish, which, when seen lying beneath the water in wait for his prey, they dexterously spear, ere the creature has time to dart with the rapidity of an arrow from the weapon about to transfix his soft but firm body."

It is this power of rapid motion, together with pelagic habits, that gives to so large a portion of the Cephalopods an extensive range in latitude, and seemingly in this direction only, for with the exception of the "poulp" (*Octopus vulgaris*), which occurs in all seas, the species of the two sides of the Atlantic are quite distinct. Though mostly pelagic, they all approach the shore at particular seasons, and some are very generally supposed to migrate periodically from south to north and back again. The successive appearance of numbers of *Loligo vulgaris, Sepia officinalis,* and the little *Sepiola* along the coast of France and later on ours would seem to confirm this notion. These habits determine the distribution of the Cephalopods over the Mediterranean area; the forms that occur there are wholly Atlantic ones, and from the western entrance to the central and thence into its extreme eastern portion, the number of species decreases progressively. As far as the *Cephalopoda* are concerned, the Western Mediterranean has fewer forms in common with the Eastern than it has with our south British seas, the reverse of what takes place as to the other Mollusca. *Philonexis, Cranchia, Loligopsis,* and *Cheiroteuthis* are amongst the more peculiar forms of the Lusitanian and West Mediterranean province. Some forms which occur on the West African coast, such as the *Argonauta hians,* have not yet been noticed within the Straits; yet this species lived in the central Mediterranean area during the later tertiary period.

The known species of Cephalopods may be taken at about 110; of these, fifty occur in the Atlantic; the "eight-armed" *Octopus,* and the "ten-armed" *Rossia, Sepia* (cuttle), *Loligo* (squid), *Onycoteuthis,* and *Ommastrephes,* are met with in its cold, temperate, and warmer regions, but the latter are richest in specific forms. *Argonauta, Philonexis,* and *Se-*

piola, mark those warmer zones which form our Lusitanian region; south of which, *Cranchia, Sepioteuthis, Loligopsis, Enoploteuthis,* and *Spirula* make their appearance.

The following tabular arrangement will serve to show both the range of certain genera in latitude, as also the general relations of the Mediterranean Cephalopods to those of the Atlantic.

	Atlantic.	Medit.	Southern Ocean.	Red Sea.
Octopus	+	+	+	+
Philonexis	+	+	0	0
Argonauta	+	+	0	0
Sepiola	+	+	+	0
Rossia	+	+	+	0
Sepia	+	+	+	+
Loligo	+	+	+	0
Sepioteuthis	+s	0	+	+
Enoploteuthis	+s	0?	+	+
Histioteuthis	+	+	0	0
Cheiroteuthis	+	+	0	0

No one has as yet undertaken the description of the testaceous fauna of the Mediterranean as a whole, though it is a work which has been long needed. It would be easy enough to compile a list of species out of the works of the several authors and observers already referred to, but this is not what is wanted: either the same eye and the same critical judgment must be applied to review the whole of the original materials collected and described by these naturalists (for some of the most distinguished amongst them differ widely in the views of specific distinctiveness), or else some M'Andrew must devote a few years to the pleasant labour of re-investigating the whole area. The first of these tasks is now hardly possible: some of the Mediterranean observers are no more, the materials they collected are already either lost, dispersed, or in hopeless confusion. The other chance alone remains.

A carefully-prepared list of Mediterranean Testacea gives more than 700 species. This is, probably, below the number. Mr. Woodward, in his excellent Manual, estimates them at 600; Mr. Jeffreys at 850: it is obvious, therefore, that this sea is wonderfully rich in this group of animals, and our knowledge of them comes nearest to that which we have of those of our own coasts. These gross results as to the Mediterranean Testacea have been obtained by summing the observations of many labourers in very many localities, some of which may be considered separately.

The Eastern Mediterranean may be divided into Northern and Southern portions. In the first of these, Risso, Payraudeau, and Michaud have collected, and our own countrymen Ed. Forbes and Jeffreys have dredged. I am unable to ascertain with what results and to what extent Ed. Forbes investigated this district; a few incidental notices, such as "dredged off the coast of Nice," are the only indications I have that he had ever worked here.

Mr. Jeffreys visited this part of the Mediterranean in 1856, for the express purpose of dredging, and with what ample spoil he was rewarded for devoting a long vacation to this pursuit along that most enjoyable of all regions—the coast of Piedmont—is fully narrated in a memoir, to which I would refer every sea-side naturalist, for his encouragement to do likewise.

In the beautiful bay of La Spezzia, the nearly tideless sea presents a very striking contrast to such as may have wandered, observed, and collected only along the Atlantic shores of Europe; throughout the whole Mediterranean, the sea-side naturalist has never presented to him any of those broad expanses of rocks and pools, swarming with life, and under so many forms, which he has been accustomed to look for where tides are lowest.

"By wading a little, however," says Mr. Jeffreys, "I found a great many live shells which I never met with in my own country, such as *Conus Mediterraneus*, and several species of *Trochus, Patella, Columbella, Vermetus,* and *Pollia*. Farther seawards is a belt or fringe of Zostera and other sea-weeds, which appears to be the favourite haunt of the *Murex Brandaris* and *M. trunculus*. Beyond this to a depth of twelve fathoms is a variety of ground, a great part being covered with Zostera and other sea-weeds; another being rocky, and the rest strong and favourable for the growth of sponges and corals." Mr. Jeffreys puts the bather on his guard against the sharp stout spines of the edible Urchin

(*Echinus esculentus*), which swarms over the rocks at slight depths from the surface.

"Outside the gulf," he continues, "is deep water; but I was disappointed in my dredging there. For several leagues seaward in from fifteen to forty fathoms, I met with nothing but tenaceous mud, with *Turritella communis* and a curious variety of *Calyptræa sinensis*. The limestone rocks of this coast contain in abundance the perforating 'Date-shells' (*Lithodomus dactylus*), by extracting which 'the fishermen eke out their precarious livelihood.'"

These few forms already quoted indicate at once to the "Celtic" naturalist that he has moved into a new zoological province. His subsequent researches will show him that all here is not equally new and strange; so that, if to the foregoing he adds *Teredina, Solenomya, Cardita, Chama gryphoides, Spondylus gaderopus, Crepidula unguiformis, Cassis Saburon, Cassidaria Tyrrhena*, and two or three species of *Mitra* and *Marginella*, he has before him the assemblage of generic difference between this sea and his own. Some of these forms are marginal, and occur readily, causing the testaceous fauna to appear more distinct from ours than it really is.

"The greatest specific variation," observes Mr. Jeffreys, "between the British Testacea and those of the Mediterranean occurs in the denizens of the littoral and laminarian zones, particularly in the genera *Mitylus, Chiton, Patella, Trochus, Buccinum, Fusus*, and *Murex*. In each of these zones certain species seem to be represented by their analogues, as *Mytilus edulis, Chiton cinereus, Patella vulgata, Trochus lineatus, Buccinum undatum*, and *Fusus islandicus* of our coasts are respectively replaced in the Mediterranean by *M. minimus, C. siculus, P. scutellaris, Tr. fragarioides, Murex trunculus*, and *Fusus corneus*.

From his own dredgings and from an examination of the collection of M. Verany, Mr. Jeffreys records as many as 375 species of Piedmontese Testacea, of which some half-dozen are new and discovered by himself; the rest are known forms, having a wide distribution either within or without the Mediterranean.

The latitude of the Gulf of Genoa is rather north of that of Vigo Bay and the north of Spain. That we may compare like things with like, Mr. Jeffreys' list may be reduced by about twenty, or to 355 species. Mr. M'Andrew's north Spanish Testacea amount to 212, giving a difference of 143 species. For the present we must take these numbers as representing these two local assemblages, and, comparing them together, we find that there are 140 species in common, leaving an excess of 215 for the Gulf of Genoa; but of these, sixty-three species are more northern and British, besides having a range down the south coasts of the Peninsula—these will be considered separately; but they reduce the Piedmontese list to

152, in which is included, therefore, the number representing the proportion of forms indicative of the province.

Of the north Spanish Testacea, which have been already noticed (pp. 108–9) as characteristically northern, some few do not seem to range much farther south, such as *Lacuna puteolus, Littorina rudis, L. littorea*, and *Purpura lapillus;* but others, such as *Velutina lævigata, Trochus tumidus, Mactra subtruncata, Tapes*, and *Pecten*, are Genoese; and what is somewhat curious, *Patella pellucida* and *Trochus cinerarius* occur on the west coast of Africa (Marocco).

Of the short list of southern shells which enter into the north Spanish fauna, some do not extend equally far north in the Mediterranean.

The bearing of Mr. Jeffreys' researches in the Gulf of Genoa on the question of marine zoological provinces, will be considered in the sequel.

Of Piedmontese Bivalves, as many as eighty are British; of the Univalves, about ninety.

"It is remarkable," says Mr. Jeffreys, "that examples of the same species are smaller than those found in the British seas. *Tellina balaustina, Jeffreysia diaphana*, and *Rissoa pulcherrima* are instances of this."

This diminution in size, which is to be observed with respect to many other species, such as *Corbula nucleus*, when traced from north to south, is the more remarkable because the converse does not take place as to southern forms in their range north. *Haliotis tuberculata*, which extends through the whole Lusitanian zone, is larger at Guernsey—which is its extreme northern limit—than elsewhere. *Ringicula auriculata* and *Mactra rugosa* are larger in Vigo Bay than in the Mediterranean, though at Vigo they are both outliers; and *Tellina balaustina*, which has its numerical maximum in the Mediterranean, is largest about the Hebrides.

With the exception of the upper extremity of the Adriatic, the sea-coast dredged by Mr. Jeffreys is the most northern portion of the Mediterranean, and it is more purely marine, for the large rivers which pour into the shallow waters of the Adriatic modify its fauna in a perceptible degree.

Mr. M'Andrew has given the results of his dredgings along a part of the Mediterranean coast of Spain (Murcia), which lies about 500 miles south of Nice, and where he obtained 353 species of Testacea. These two local assemblages admit of comparison; their numbers, omitting Mr. Jeffreys' new species, are very close, and the portions of coast examined were of about the same extent. The result is that the two sets are nearly identical; the specific difference for the most part hardly deserves notice. *Cymba olla* reaches up the Spanish coast as far as Malaga, so that it has a Mediterranean range corresponding to its Atlantic one. *Solarium luteum* and *S. stramineum* reach so far, but have not been observed farther, whereas on the Atlantic side they extend to the north coast of Spain.

The seas of the Two Sicilies are wonderfully prolific in animal life, and it is only in this region that we have opportunities of comparing the products of localities with what they were 2000 years ago. Tarentum was the resort of the Roman epicures; its abundant fish-markets gave it half its bad reputation, encouraging its wealthy wool-dyers and clothiers to indulge in fish dinners of profuse extravagance. The proud boast of the Tarentines, "that others prepared by labour for the future, but that they, by means of their banquets, were not about to live, but were already living," was based on the products of its inner and outer seas.

"Pectinibus patulis jactat se molle Tarentum."

The great inland bay, the "mare picolo," swarms with scallops now as it did in the days of Horace. It still affords the main support of the fishing population. Mounds of pounded shells mark the sites of the old dye-works of Tarentine purple; indeed, its old trade still lingers on. There the Murices are as abundant as ever, so are the Mullets and the Tunnies.

The Syrian and Tarentine purples were of several tints, and the shell-fish employed were of two sorts at least; one of these was certainly the *Murex trunculus*, which is most abundant here in the marginal zone, and indeed throughout the whole of the Mediterranean.

This copious marine Sicilian fauna has been fully and even magnificently illustrated: there are, first, Poli and Delle Chiaje; more recently, Cantraine and Philippi; nor must Milne Edwards be forgotten, who, in his Sicilian researches, put on the helmet of the submarine diver, and passed whole hours in collecting and observing beneath the clear waters of that sea.

For all purposes of numerical and other comparisons the work of Philippi is the best guide as to the Mollusca of the central Mediterranean; he there enumerates 522 species of Testacea, and the whole assemblage is presented in its relations to the older fauna of that area, as exhibited in the tertiary strata of Italy and Sicily, as well as with those of certain remote faunas of the present period.

If Philippi's 522 species are taken as a fair and typical representation of the Mediterranean Testacea, and the species which are both recent and fossil, amounting to 360, be deducted, the remainder, 162, gives the difference between the present fauna and a former one.

Comparing this typical assemblage with our own, we have

	British.	Sicilian and Italian.	Difference.	Common species.
Bivalves	156	188	32	83
Pteropods	4	13	9	
Gasteropods	232	313	81	57
Cephalopods	14	15	1	7

Philippi took the works of Fleming and Montague

as the basis of his comparisons, and estimates the Bivalves at 198, the Brachiopods at 5, the Gasteropods at 191; in all 394 species. If we take the more critical and newer enumeration of the "British Mollusca," the same orders give 397. This close agreement after an interval of so many years, and after so much research, is somewhat remarkable, and it is only on a careful examination of the species composing these numbers that it is seen what a great change our British list of Mollusks underwent in the hands of Forbes and Hanley.

If we add to Philippi from other sources and enumerate the central Mediterranean Bivalves at 200 species, one-half will be also British, an amount of agreement sufficient to indicate the Atlantic character of the fauna, when it is remembered that the comparison is made between the denizens of zones separated by 5° of latitude. These two zones extend respectively from 35° to 45° N. L., and from 49° to 59° N. L., making their extreme limits 14° apart. If, however, the British list is reduced by some forty species, which are northern in their British range, there remain only twenty-three species to characterize our south-western assemblage, or a rate of change of not more than from three to four species for each degree of latitude, when compared with the Mediterranean. Bivalves have a broader and more uniform distribution than other classes of their order.

Comparing the enumeration given by Philippi

with those already cited of Jeffreys and M'Andrew, the agreement is found to be very close, as might be expected.

One other numerical comparison will suffice. Ed. Forbes' observations in the Eastern Mediterranean are so much fuller than those of the French naturalists on the coasts of Greece that his enumeration of the Testacea is the best that can be taken. He there procured of Gasteropods 254, Brachiopods 8, and Bivalves 242, in all 494 species; the Sicilian seas, under the same classes, giving 475.

There is a very great amount of agreement between the Ægean and the Sicilian lists of Testacea, more particularly when the comparison is made between the denizens of the higher sea-zones; and the main difference exists with respect to the inhabitants of those deeper regions to which Philippi had no means of access.

Of all local assemblages that which Ed. Forbes has given for the Eastern Mediterranean is probably by far the most complete. From the uniformity which prevails at depths, both as to conditions and distribution, the deep sea forms of the Ægean may be supposed to occur equally in the Sicilian seas; in other words, the Eastern Mediterranean must be somewhat poorer than the central and Western portions.

" The absence of certain species in the Ægean, which are characteristic of the Western Mediterranean, is rather to be attributed to sea-composition than to climate," says Ed. Forbes. This he thinks is due to the pouring in of the waters of the Black Sea. This influence is uniform over the whole of the Eastern Mediterranean, and has been stated by Ruprecht to be appreciable on the Syrian coast.

Differences based on negative evidence are never safe supports for any inferences; indeed, unless the amount of such evidence is very considerable, such differences are hardly worth noticing. Comparing what is known of the Molluscous fauna of the extreme Eastern and Western portions of the Mediterranean, all those species procured by Ed. Forbes from depths in the Ægean, such as have not been dredged in other parts of that sea, must be carefully excluded.

The Mediterranean fauna, however, being so essentially Atlantic, it may reasonably be expected, seeing the great eastern extension of this internal sea, that it should present a certain amount of decrease in that direction. A marine fauna which is an offshoot of another must not be considered as the definite result of one migration, but as an assemblage which has been constantly modified by the slow extension of species.

The reputed peculiarities of the eastern-division Mollusks are *Clavagella balanorum, C. angulata*, and *C. Melitensis, Thecidea Mediterranea, Umbrella Mediterranea, Murex cristatus, Pedicularia sicula, Dolium galea, Cassidaria Tyrrhena*, and *C. depressa, Trochus Sprattii, Venerupis decussata, Pecten Jaco-*

bæus, &c. Had this local assemblage been more numerous it would have been interesting to have traced its extra-Mediterranean relations ; but, small as it is, it would not be without its value if all the species pointed to some common province or locality ; but such is not the case. Two of the *Cassidariæ* are old occupants of the great Mediterranean basin; and though *Dolium galea* is found in the Red Sea, it by no means follows that it came from thence, inasmuch as, together with *Umbrella*, it has a wide distribution in the South Lusitanian Atlantic Province, and is also one of the old Mediterranean fossil forms.

From some observations made in the neighbourhood of Algiers, it was found that through all the months of the year the temperature of the water decreased from the coast-line outwards, as also from the surface downwards; this decrease is greater in summer than in winter. The temperature of the water is higher than that of the air in autumn and winter, lower in spring and summer. In the deeper zones it falls as low as 54° F., which it never passes, as has been ascertained for depths of from sixty to 360 fathoms.

The mean winter temperature of Toulon is 52° F., that of Algiers 56° F., the mean being 54° ; in the Adriatic the mean temperature of the air is between 59° F. and 73° F., that of the water being between 66° F. and 71° F. The difference is not great, but, so far as temperature can influence the extension of species, southern forms find a more congenial one on the Algerine coast than elsewhere. If to this condition is superadded that of the inflow from the Atlantic, setting along the North African coast, there exist good reasons for expecting certain local peculiarities.

The Atlantic inflow determines that marked preponderance of pelagic animals which is to be noticed about the Gut of Gibraltar. The forms of Mollusks which are western are such as *Ervillia castanea, Siphonaria concinna, Acmœa virginea, Mesalia sulcata* and *M. striata, Cymba olla, Lutraria elliptica, Venus striatula, Astarte sulcata* and *A. triangularis, Natica intricata;* these are all Lusitanian and West African species. In addition, these forms and some others are not found fossil in any of the raised sea-beds of the older Mediterranean, and may therefore be looked upon as species which as yet have made only a limited progress in colonizing this internal sea.

The Eastern Mediterranean is inseparably connected with Edw. Forbes' researches into the distribution of animal and vegetable life in depth; he found, proceeding from the highest upward limit of these waters downwards, that there were the following distinct zones.

The Littoral Zone has a depth of only two fathoms; and, small as this is, it yet admits of a twofold division, even in this nearly tideless sea. The narrow interval between tides is thus described by Ed.

Forbes :*—"The testaceous Mollusks of the shores of Lycia are numerous, but are more remarkable for variety than for their dimensions. On the rocks near the water's edge, *Patella scutellata* and *Bonnardi* are common; also the *Haliotis lamellosus* and *Fissurella*. Under stones near the water-mark *Chiton siculus* is abundant; more rarely, *C. fascicularis* and *Cajetanus*. *Littorina petræa* is found at the very edge of the water, not differing from specimens from the west coast of Britain. Species of *Vermetus* indicate the zoological character of the province; also numerous forms of *Trochus*, of which the *T. Lyciacus* has not been observed elsewhere. *Murex trunculus, Pollia maculosa, Columbella rustica, Fasciolaria Tarentina, Fusus lignarius*, and *Conus Mediterraneus*, all shells of handsome aspect and sub-tropical forms, are abundant in similar situations. Under the large stones in the little creeks is found one of the larger European forms of Cowrie (*Cypræa spurca*). Many curious Bivalves live attached to the rocks along the coast-line, or in their crevices, such as *Cardita calyculata, Arca barbata, Spondylus gadæropus, Lima squamosa,* and the Date-shell (*Lithodomus lithophagus*).

"Where the coasts are of sand we have a different set of Mollusks. Immediately along the water's edge, at a depth of an inch or so beneath the sand, are buried myriads of a little Bivalve—*Mesodesma donacilla*. *Solecurtus strigillatus* is found farther

* Travels in Lycia, vol. ii. p. 102.

out, and buried deeper; also *Lucina Desmarestii, Amphidesma sicula,* and the curious *Solemya Mediterranea.* Where the sand is coarse, *Venus decussata* is found.

"On muddy shores *Lucina lactea* abounds, and where a stream pours in may be seen millions of *Cerithium mamillatum,* along with some minute *Rissoœ.*

"*Mactra stultorum, Kellia corbuloides, Lucina pecten, Venerupis decussata, Donax trunculus, Cardium edule, Emarginula huzardi, Truncatella truncatula, Cerithium fuscatum, Nassa neritia* and *gibbosula,* and *Auricula myosotis* complete the list of the most constant Molluscan inhabitants of the Lycian shores to a depth of seven or eight feet."

The whole number of Mollusks referred to this region, consists of thirty-eight species of Bivalves and 109 Gasteropods; and of these two classes, twenty-seven and fifty-two may be taken as reprerenting the proportion of species which have their maximum in this zone.

Among the Zoophytes, the littoral rocks of the coast of Lycia are distinguished from those of the Ægean islands, by masses of *Cladocera cœspitosa,* never living deeper than eight feet from the surface: large sponges grow in the sheltered gulfs, and *Padina pavonia,* which has an Atlantic distribution as far as our own southern shores, is the characteristic plant of this zone.

Such is the character of the littoral fauna of

the Eastern Mediterranean, as it is presented to the eye of the casual observer. Storms wash up the shells which belong to the lower portion of the marginal belt, and the whole goes to form the assemblage which is to be commonly met with along its shores. If, however, the sea-side naturalist from this country should neglect these dead spoils, and confine himself to such living forms as may be collected over the narrow belt between land and water, he will meet with little to remind him of his southern latitude. Of the eleven species of Mollusks peculiar to this upper belt, eight have a wide Atlantic distribution; he will collect *Littorina cœrulescens, L. petrœa, Kellia rubra, Truncatella truncata*, as he might on our own coasts; also our common Barnacle; and in addition to *Padina*, such plants as *Dictyota dichotoma* and *Corallina officinalis*, in wonderful profusion.

The general assemblage of forms, which imparts a sub-tropical aspect to the coasts of the Mediterranean, is derived wholly from depths of a few feet below the permanent sea-level.

The Second Region extends from two to ten fathoms. With a sea-bed of sand or mud, the former is usually covered with the beautiful green *Caulerpa prolifera*, the latter with "grass-wrack;" other sea-plants abound. The characteristic Testacea of this zone are—*Pecten polymorphus, P. hyalinus, Tellina donacina, T. distorta, Modiola, Nucula margaritacea, Lucina lactea, Cardium exiguum, C.*

papillosum. Among the *Gasteropoda, Cerithium vulgatum* and *C. lima* are most abundant. *Trochus crenulatus, T. Sprattii, Rissoa ventricosa, R. oblonga, Marginella clandestina; Pleurotoma,* four species; *Natica olla; Phasianella,* three species; *Nassa,* six species; and *Mitra obsoleta*. The Stony Coral (*Caryophyllia cyathus*) commences in this zone, and ranges down through all succeeding ones.

The Bivalves are fifty; twenty-two have their maximum.

The Univalves are seventy-six, sixty-two have their maximum.

The Third Region reaches from ten to twenty fathoms, with a sea-bed of gravel or sand; the same plants are continued, but become scarcer towards the lower limits of the region; a change has become obvious at depths of 160 feet, but the upper limit is not so distinct or definite. The Testacea, also, are to a very great extent the same.

The Bivalves of this zone are fifty, of which seventeen have their maximum here.

The Univalves are under seventy, and twenty-six are at the maximum here.

Bivalves predominate numerically.

The Fourth Region extends from depths of twenty to thirty-five fathoms. Here, a great part of the species of the upper zones are replaced by others, which are curiously representative of them in form. This takes place in still lower regions. There is no transmutation of one into the other; each has all its characters precisely defined, and usually before the characteristic species of one region has declined to the numerical minimum of individuals, its successor has appeared; at first scarce, but when in its true or proper region, as abundant as its predecessor had been. The characteristic Fuci are *Dictyomenia volubilis, Sargassum salicifolium, Codium bursa, C. flabelliforme,* and *Cystoceira*. The rare and curious *Hydrodictyon umbilicatum* was procured in this region. Urchins are abundant, and *Comatulæ*.

The Bivalves are about sixty, of which thirty-three have their maximum here.

The Univalves are about ninety, with about forty having their maximum here.

Terebratula detruncata and *cuneata* make their appearance here, the latter being most abundant.

Retepora cellulosa abounds; and *Myriapora truncata,* with *Cellaria ceramioides,* are characteristic. The finest sponges of commerce are taken from this zone. *Nucula emarginata* serves generally to mark this depth.

The Fifth Region extends from thirty-five to fifty-five fathoms; its plants are *Rytiphlæa tinctoria* and *Chrysimenia uvaria. Dictyomenia volubilis,* which gives a marked character to the preceding zone, becomes scarce here. The sea-bed is composed of Nullipores, and is shelly. The Testacea most generally distributed are *Pecten opercularis, Turritella tricostata,* and the most abundant in individuals are *Nucula emarginata* and *striata*, *Cardium papillosum*, *Cardita aculeata*, and *Dentalium novemcostatum*.

Echinoderms are frequent here; not so Zoophytes.

Terebratulæ increase. This is the proper zone of *T. detruncata* and *Crania ringens; T. seminula* makes its appearance.

Bivalves fifty-six; at their maximum, twenty-five. Univalves seventy-six; at their maximum, thirty.

The Sixth Region reaches from fifty-five to seventy-nine fathoms. The sea-bed is here covered with Nullipores; and Fuci are extremely few. Though such is the case, there is still found here a considerable number of Phytophagous Mollusks, and which feed on the vegetable Nullipores.

The Testacea which are most generally diffused are *Venus ovata*, *Cerithium lima*, and *Pleurotoma Maravignæ;* those most prolific are *Turbo sanguineus, Emarginula elongata, Nucula striata, Venus ovata, Pecten similis,* and the various species of Brachiopods.

Below fifty fathoms, the range of each set of Mollusks extends wider as we descend.

Cidaris hystrix is the characteristic Echinoderm.

The Bivalves are forty-six, with seventeen at their maximum. The Univalves about forty, with twelve at their maximum.

Terebratulæ increase, in addition to those of the higher zones. *T. truncata* and *T. cuneata* appear, with *Crania*.

The Seventh Region, ranging from eighty to a hundred and five fathoms, has characteristic features. Herbaceous Fuci have disappeared, and Nullipores are the only plants. The Tunicated Mollusks have ceased, as also have *Nudibranchiata*. Of Testacea, *Lima elongata*, *Cardita aculeata*, *Rissoa reticulata*, and *Fusus muricatus*, are most generally distributed, and the same *Rissoa*, with *Turbo sanguineus*, *Venus ovata*, *Nucula striata*, *Pecten similis*, together with the Brachiopods, which abound here, are the most prolific. Echinoderms are not uncommon, such as *Echinus monilis*, *Cidaris hystrix*, *Echinocyamus pusillus*, and some *Ophiuridæ*, but no *Asteriadæ*.

The Bivalves of this zone are under thirty, with about nine at their maximum. The Univalves are about forty, with sixteen at their maximum. In addition to all the fore-cited *Terebratulæ*, are *T. lunifera*, *T. vitrea*, and *T. appressa*.

The Eighth and lowest region includes all depths below a hundred and five fathoms. Over this seabed, which consists of a fine yellow sedimentary mud, full of the remains of Pteropods and Foraminifers (an unknown region till the researches of Ed. Forbes and his associates in the Beacon), there is found a uniform fauna distinguished from all preceding regions by peculiar species. As would be expected, this region gave a large proportion of new species, such as *Pecten Hoskinsii, Lima crassa, Nucula Ægeensis, Scalaria hellenica, Parthenia fasciata* and *ventricosa.*

Ligula profundissima, Pecten similis, Arca imbricata, Dentalium quadrangulare, and *Rissoa reticulata,* are more prolific in individuals in this region than in any other.

Ophiura abyssicola, Amphiura florifera (Chiaje), and *Pectinura vestita* are the Echinoderms of this eighth region, and are well fitted, by their organization, for living in the mud of these depths. *Caryophyllia cyathus, Alecto,* and *Idmonea* range down to these depths.

Ligula profundissima and *Dentalium quinqueangulare* are the most generally diffused species below 105 fathoms. *Nucula, Neæra, Arca,* and *Kellia* live as deep as 180 fathoms. *Arca imbricata* was taken alive as low as 230 fathoms.

The *Terebratulæ* would seem to have their limit in the seventh region.

Of these eight regions of depth of the same sea, the Ægean, the highest and the lowest have only two species of Mollusks in common—*Arca lactea* and *Cerithium lima,* and there is a doubt whether the last may not be a straggler from the zone above. Regions three to eight inclusive have only two species in common.

These results are of such significance to the Palæontologist, that for his use the following table of Testacea, compiled by Ed. Forbes, is here reproduced.

	2 fath. i.	10 fath. ii.	20 fath. iii.	35 fath. iv.	55 fath. v.	79 fath. vi.	105 fath. vii.	230 fath. viii.
Chitons	7	3	2	0	2	2	1	1	0
Patelliform univalves	20	11	3	2	3	5	6	6	1
Dentalia	6	4	4	2	2	1	1	2	2
Spiral univalves, Holostomatous	115	50	40	40	44	35	28	17	15
Spiral univalves, Siphonostomatous	104	40	27	30	41	36	30	16	5
Pteropods and Nucleobranchs	12	1	0	0	0	0	0	3	12
Brachiopods	8	0	0	0	2	4	5	7	3
Lamellibranchs	135	38	53	52	68	58	48	34	28
Ægean total :	407	147	129	126	162	141	119	86	66

The system of representation in depth, which has been alluded to in the notice of the fourth region, is of this kind. Representative forms are similar, and require the tact of an experienced naturalist to discriminate them. One or two examples will suffice in order to show how such forms replace one another in zones of depth.

		i.	ii.	iii.	iv.	v.	vi.	vii.	viii.
Nucula	emarginata	—	min.	—	max.	min.	—	—	—
	striata	—	—	—	min.	max.	min.	-	—
Arca	barbata	max.	—	—	—	—	—	—	—
	lactea	min.	—	—	max.	—	—	—	min.
	scabra	—	—	—	min.	—	—	max.	min.
	imbricata	—	—	—	—	min.	—	—	max.
Trochus	crenulatus	—	max.	min.	—	—	—	—	—
	exiguus	—	min.	—	—	max.	—	min.	—
	ziziphinus	—	min.	min.	max.	—	—	min.	—
	millegranus	—	—	—	—	min.	—	max.	min.

A like system of representation is to be traced, as will be seen, amongst the forms found in the older sedimentary deposits.

Reference has been already made to the species of Testacea common to the Mediterranean and our own seas. In the bathymetrical distribution of species, those which have the most extensive range in depth have also a very wide geographical distribution. Thirty-eight species range through four out of the eight regions of the Ægean. Of these at least twenty-one are British, and hence Ed. Forbes arrived at the general inference that *"the extent of the range of a species in depth is correspondent with its geographical distribution."*

The distribution of Celtic forms in the several Ægean zones has been represented in the following tabular form:—

	i.	ii.	iii.	iv.	v.	vi.	vii.	viii.
Chitons	1	0	0	1	1	1	1	0
Patelliform univalves	0	1	2	2	2	2	2	0
Dentalia	1	1	0	0	0	0	0	0
Spiral univalves, Holostomata	12	9	13	16	14	11	8	4
Spiral univalves, Siphonostomata	4	5	7	8	9	6	5	2
Pteropods and Nucleobranchs	0	0	—	—	—	0	0	0
Brachiopods	0	0	—	0	—	0	0	0
Lamellibranchs	16	25	28	39	33	19	11	7
Total	34	41	50	66	59	39	27	13
Percentage	21	36	45	43	40	35	36	20

Forms decrease in size from higher to deeper regions, and in the same direction they part with all their brilliant colouring and variety of pattern. Well-defined patterns are, with few exceptions, presented only by Testacea inhabiting the littoral and median zones. " In the Mediterranean one in eighteen only of the shells undoubtedly belonging to depths of 100 fathoms and upwards exhibited any colour-markings, whilst the proportion of the coloured to the colourless, from between thirty-five to fifty-five fathoms, was as one to three."

In our own seas the same species which are even vividly coloured and banded in higher zones are colourless when taken from depths below 100 fathoms; a like absence of ornament takes place proceeding northwards, at depths even of sixty fathoms.

Such is a brief and general outline of the change which takes place in the Eastern Mediterranean, from the surface to depths of upwards of 1300 feet. Any lengthened enumeration of specific forms has been purposely avoided. The features which are noticed are only those broader ones which are derived from positive characters. Each zone has a distinctive sea-bed, with certain peculiar forms. As we descend the dimensions of each zone become greatly extended, so that whilst the upper has a depth of only twelve feet, the lowest ranges through 700 feet. Specific animal forms decrease rapidly, and just as the sub-aerial zones of vegetation present us at last, as we ascend, with only such forms as lichens, so at depths of from seventy to a hundred fathoms, we have the obscure Nullipores as the extreme forms of marine vegetation.

The Mediterranean fishes are so well known and have been so admirably illustrated, that we have little difficulty in comparing them with those of the Atlantic. This state of our knowledge is owing to the vast importance of the fisheries of this sea to the dwellers around its shores, so that the habits and migrations of many of its valuable species as articles of food, had been accurately observed and recorded as far back as 2000 years since. From those more recent days when Natural History became a pursuit and a study, the peculiar beauty of some of the Mediterranean fishes could not fail to attract attention. Even the least observant of those whom this country sends forth annually, to wander along the shores of southern Europe, can hardly have failed to notice the striking difference between the contents of Italian or Sicilian fish-markets and our own—a contrast as great as that which Nature there displays in all her other aspects.

Risso estimated the fishes of the Eastern Mediterranean at about 400, and though his enumeration cannot be implicitly accepted, yet, after making all deductions, the additional species to be derived from the great work of Cuvier and Valenciennes, again more than bring up the number (p. 20). Such a list is far larger than any for which we have certain data with reference to the Atlantic portion of the Lusitanian province ; as, however, we possess the list of our own British fishes on one side, and that of the Atlantic islands on the other, forming the northern and southern limits of that province, the special relations of the fishes of the Mediterranean can be easily determined.

If we take our British fishes at 270 species, we shall find that no less than 150, or upwards of thirty-seven per cent., are also Mediterranean ones. It must not, however, be supposed that the species common to that sea and ours are uniformly distributed ; there is within our area a very distinct limitation of forms both northern and southern, and even eastern and western ; so that the comparison with the Mediterranean will be close or otherwise, according as we include or exclude our southern and western fishes. If we take the Cornish fishes by themselves, as they may be collected during the summer season from out of the Mount's Bay fishing-boats, the Mediterranean aspect of the assemblage will be very striking ; excluding the north Celtic forms—such as occur but rarely, if ever, on our south-western shores—the agreement would amount to nearly one-half.

Our imperfect acquaintance with the ichthyology of a large portion of the Lusitanian province has already been noticed (p. 118), so that it was found necessary to travel as far as the extreme southern limits of the province for any list of well-determined species, and it was there shown that the Madeira fishes occupied a somewhat intermediate place between our assemblage and that of the Mediterranean, and that they were less tropical in their aspect than the latitude of the island would lead us to expect.

There is a peculiarity with reference to these Madeira fishes, which is still more striking when those of the whole of the Canaries are taken together, and which may be noticed here. Those diligent collectors, Messrs. Webb and Bertholet, submitted to M. Valenciennes upwards of 110 species of fishes from those islands; amongst these are to be found such forms as *Priacanthus boops*, *Beryx decadactylus*, a genus poor in species; *Pimelepterus incisor*, *Caranx analis*, and *Coryphœna equisetis*. With some few exceptions of this kind, namely, of common forms which serve to connect the two sides of the Atlantic, the fishes of the Canaries are mainly such as are also met with in the Mediterranean; with this difference, however, that certain species which are scarce in the Mediterranean are common about the Canaries, and of these many range down the African coast as low as the Cape, and also in advance of it—as to Ascension and St. Helena.

When the extended migration of certain fishes is considered, it will not, perhaps, be thought strange that the two sides of the Atlantic should have some few forms in common, even as low down as between the Canaries and the Brazils; but it is well observed by M. Valenciennes, that this community has reference in some cases to fishes which are not migratory, such as *Pimelepterus incisor;* and that the form and nature of a coast-line influence the distribution of fishes far more than temperature dependent on latitude. The migratory movements of fishes, like those of birds, are made in obedience to given wants and instincts, and are conducted, like the voyages of the early navigators, not across the trackless depths of the ocean, but along lines of coast.

It is somewhat remarkable, that so far as our present knowledge goes, some of the American forms of fishes found about the Canaries do not reach the African coast, a consideration which, in conjunction with others we shall have to notice, lends support to the view already put forth (p. 113), that the western boundary of the "old world" was once placed so much farther westward, as to reach these Atlantic islands. The occurrence of common forms, more particularly of the species which have been cited, can only be explained by the coasts of the two sides of the Atlantic having once been placed much nearer to one another along some line south of our Lusitanian province.

Of the fishes of the Canaries, seventy are found in the Mediterranean; of these, many are also West African forms. There are some others, such as *Pristopoma ronchus, Sargus cervinus, Chrysophrys cæruleosticta,* and *Lichia glaycos,* which are also West

o

African, but which have not as yet been observed within that internal sea. These, like the American forms, have to be deducted when the fishes of the external Lusitanian province are compared with those of its Mediterranean portion. The known fishes of this sea are 444, and of these as many as 150 at least range the Atlantic as high as our south-western coasts, and seventy are met with about the Atlantic islands. If we compare the numbers of the Mediterranean fishes with our own (p. 119), we find that an interval of twenty degrees of latitude across the whole Lusitanian province gives still as much as twenty per cent. of common forms. This is a very low rate of change in space with respect to the distribution of species; but it is quite high enough to satisfy us (in the absence of any definite information respecting the fishes of the Atlantic coasts of Spain and Portugal) that, so far as calculation can be applied to such questions, there exists a very high degree of probability that all the known species of Mediterranean fishes are also Atlantic.

It is but justice to Risso, the naturalist of the seas of Nice, that he should be mentioned as perhaps the first who called attention to that distribution of marine life which is dependent on depth. With reference to the great class of fishes, he observes that certain forms frequent the mouths of rivers; that along the open coast-line the submerged marginal rocks, covered with *Fuci, Cera-*

miæ, and *Confervæ*, are the haunts of Blennies, Dragonets (*Callyonomus*), Gobies, Pipe-fishes (*Syngnathi*), and Sea-snipes (*Centrisci*). The shelving beds of shingle and sand are the zone of a numerous tribe, such as Launces (*Ammodytes*), *Lepidogaster*, Garter-fishes, *Lepidotus*, *Labri*, Wrasses (*Crenolabri*), the Sea-breams (*Sparus*), Smelts (*Osmerus*), *Gymnetrus*, *Scopelus* (always found in company with the Anchovies), Sardines (*Clupanodon*), and Mullets.

The region of *Algæ* and *Cauliniæ* is that of the Denzelles (*Ophidium*), *Murænæ*, Star-gazers (*Uranoscopi*), and of the Scorpions (*Scorpænæ*). At depths of twenty-five fathoms or thereabouts, or over the zone of *Bryozoa* and Zoophytes, are the File-fishes (*Balistes*) ; also the genera *Chauliodus*, *Murænophis*, *Labrus*, *Dentex*, *Lichia*, *Peristidium*, and certain Gurnards. A muddy sea-bed, with a depth of fifty fathoms, is the favourite abode of the Ray tribe, of the Angler (*Lophius*), the gigantic *Cephaloptera*, and of the Plaice. At like depths, but without any special reference to sea-bed, are found Whitings, Cod, *Holocentrus*, *Citula*, *Seriola*, *Tetragonurus* (the "Corbeau" of the Mediterranean), and certain species of *Sparus*.

Lowest of all come the *Alepocephali*, of which Risso remarks that, in common with other fishes taken at depths of 2000 feet and upwards, it has its scales very feebly attached to its skin, the eyes disproportionately large, a large swimming-bladder, and numerous cæca. They also present few tints. *Pomatomus* is taken at 250 fathoms. *Chimæra* and *Lepido leprus* also belong to this zone.

During the researches of the "Beacon" on the Lycian coast and among the islands of the Ægean, above seventy species of marine fish were observed, examined, and drawn, being more than twice the number recorded from the Grecian seas in the great French work on the Morea. Fishes are numerous in the Eastern Mediterranean, but very few attain any considerable size. "In the sheltered bays and gulphs are numerous species of *Sparoideæ*, a tribe very characteristic of this region; forms of *Sargus, Pagrus, Chrysophris, Cantharus, Sparus, Dentex, Boops,* and *Oblada*. They may be seen swimming in shoals around the vessels at anchor, their broad, silvery sides glancing in the water, in some striped with irregular bands of gold, in others marked with one or two dusky clouds, or tinged with brilliant ultramarine and purple. They are abundant in water from five to seven fathoms deep, where the bottom is muddy or weedy. The *Scarus creticus* is abundant on the Lycian shores: it is remarkable for the variation in colour it presents at different seasons, at one time being of the most vivid crimson, at another of a dull bluish-grey, and sometimes piebald of the two colours. Equally and even more vivid are the Wrasses, of which many gorgeous sorts are common among the rocks close to the shore. The *Julis Mediterranea* is the brightest of these painted beauties, exceeding all fishes of the Mediterranean for splendour of colour. Some of the species of *Sphyræna* glow with the brightest vermilion. These usually replace the Wrasses, being found in deeper water."

Immense flocks of the little *Atherina presbyter* may be seen on fine days skipping on the surface of the water, endeavouring to escape from the needle-like Gar-pike. There is a great Grey-Mullet fishery carried on in Caria. The Red Mullet (*Mullus barbatus*) is everywhere abundant. In sandy creeks the *Uranoscopus* is frequent. Species of Sole and other flat-fish, the Torpedo, of which *Torpedo narke* is the most frequent, also occur in similar situations. In rocky nooks, besides the beautiful Wrasses, Blennies and Gobies abound, some of them brilliantly coloured. Under great masses of rock close to shore lives the *Muræna*, its long, slimy body beautifully clouded with purplish-brown and salmon-colour. The fish which was found to live deepest in the Ægean was a little Goby, which was frequently taken in the dredge at a depth of forty or fifty fathoms.

Sea-Turtles are such exceedingly rare visitors to our Celtic latitudes, or indeed into the northern Lusitanian, that their occurrence in the Mediterranean becomes one of the characteristic features of the fauna of that sea. If to these forms are added the fresh-water Tortoises, which abound in the low circumlittoral lakes and marshes of this region, as also the Crocodiles and Monitors of its African boundary, we see how largely, and with what striking forms, the Reptilian order becomes represented here.

The Leathery Turtle (*Sphargis coriacea*) ranges throughout the whole extent of the Mediterranean, and even into the Black Sea. In the waters of the eastern portion it is common, and it was of the shell of this species that, according to the mythology of Greece, the first stringed instrument of music was made.

Its breeding-places are along the sandy shores of the southern Mediterranean; it is here that it is most abundant, and it is rather western than eastern. It attains a great size—upwards of seven feet in length; its paddles are long and broad.

Somewhat rarer than the foregoing is the Caouane (*Testudo caretta*), also southern and western in its Mediterranean range. It resorts to the coasts of the island of Sardinia, where, as at Cagliari, considerable numbers are taken. Both these Turtles have an extensive Atlantic distribution, reaching far down the West African coast, and across to those of America; they are true members of the Mediterranean fauna, but represent its West African, rather than its Lusitanian elements.

The Mediterranean has no peculiar Cetaceans. The Atlantic forms which ordinarily range there are few, and have been mostly recorded from the Gulf of Genoa and the western portion.

The common Dolphin (*Delphinus delphis*) of our

classical associations, the emblem of so many of the old Mediterranean States and cities, is still the most common species. The great Dolphin of the Atlantic (*D. tursio*) only occasionally finds its way there. The so-called Mediterranean Rorqual (*Balænoptera musculus*) is Lusitanian and Celtic ; it is only more Mediterranean because, like the common Dolphin, it has a less northern Atlantic range than certain others of its order.

Some early notices would lead us to suppose that the *Cetacea* may formerly have ranged more freely over the whole length of the Mediterranean than they do at present. Such also appears to have been the case as to the Lusitanian and Celtic portions of the Atlantic.

Of the amphibious *Carnivora*, the common Seal (*Phoca vitulina*) ranges down from northern latitudes into the south, and enters the Mediterranean ; but it is doubtful whether it is amongst the species found in the Black Sea and the Caspian. The Adriatic Seal, "the Monk" (*Pelagus monachus*), so abundant about the islands of the Dalmatian Archipelago, and the fiords of that solitary coast, is also the common Seal of the Grecian seas. This is a sub-genus, founded on dental characters, and of which the form in question seems to have an Eastern and somewhat limited Mediterranean range.

CHAPTER VII.

THE BLACK SEA.

THAT large area which is comprised within the island of Crete and the shores of Greece, and Asia Minor, is in most striking contrast with all other parts of the Mediterranean basin. Viewed as an area of depression, the history of this region is probably the same as to date with all the rest ; but if so, its original features were very distinct : lofty islands, rocky coasts, and deep intervening seas form for 350 miles the approach to the narrow straits which lead into the Black Sea. A glance at a good physical map of this region will suffice to indicate that the islands of the Ægean are the peaks and ridges which once connected the mountains of Greece with those of Anatolia.

The opposite shores of the Dardanelles and Bosphorus approach so close at places as to give to this connecting link between the Ægean and the Black Sea the features of a broad river ; and this resemblance is increased by the steady flow of the water outwards. This "set" of the "ocean stream" may be observed in parts of the Ægean ; it is the excess of inflow into the Black Sea beyond the loss by evaporation.

The Black Sea, from east to west, is about 700 miles in length, with a breadth of 300, giving an area of 170,000 square miles; its depth in places is nearly equal to that of the Mediterranean. Beyond this again is another expanse of water—the Sea of Azof, with an area of 14,000 square miles; this is a shallow sea.

Into these two depressions, which together exceed the area of the British Islands, some of the largest rivers in Europe discharge themselves, such as the Danube, the Dnjepr, the Dnjestr, the Bug, and the Don. To these might be added an almost endless list of minor rivers, many of which far exceed the volume of the largest British streams.

Some of the rivers which discharge into the Black Sea take their rise in high latitudes, in districts annually covered with snow. These rivers also are annually frozen. Again, the winter temperature of the northern shores of this sea is such that coast ice forms there, as also in the Sea of Azof; and hence the waters of the Black Sea are much colder·than those of the rest of the marine province to which it belongs. It is to the combined influence of composition and temperature that the great difference in the assemblage of animals in the Mediterranean and Black seas must be attributed. The Black Sea is the great ultimate estuary of the rivers which drain one-half of the European area.

The proportion of Baltic Testacea to those of the Celtic Atlantic region is as fifteen to three hundred.

The Black Sea species are to the Mediterranean as sixty to six hundred ; of these about thirty, or one-half, are British. The Molluscous Fauna of the Black Sea is Atlantic, and the assemblage of species, as well as their relative frequency, causes it to resemble the northern portion of the Lusitanian zone. The species common to our seas and the Black Sea are *Cerithium adversum, Littorina rudis* and *neritoides, Trochus umbilicatus, T. cinerarius, T. exiguus, Phasianella pulla, Calyptræa chinensis, Murex erinaceus, Nassa reticulata* and *ascanias, Anomia ephippium, Cardium edule* and *exiguum, Venerupis irus, Venus aurea, V. gallina, V. dysera, Tellina tenuis, T. carnaria, Mactra triangula, Solen ensis, Pholas candida.*

The Lusitanian or Mediterranean species are, *Patella tarentina* and *ferruginea*, three or four species of *Rissoa, Truncatella truncatula, Cerithium vulgatum, C. ferrugineum, Trochus divaricatus, T. Adansonii, T. villicus, T. fragarioides, Pleurotoma costulatum, Tritonium corniculum, T. neriteum, Columbella rustica, Conus Mediterraneus, Bulla striata.* The bivalves are *Ostrea Adriatica, Pecten sulcatus, Mytilus minimus, M. latus, Lucina commutata, L. lactea, Venus rudis, Mesodesma donacilla, Erycina ovata.* These two lists convey a very fair representation of the assemblage of the Black Sea Mollusks. A few more species might be added. All that are here cited rest on the careful identifications of Dr. A. Von Middendorff, and the peculiarity of the assemblage of marine species consists in the dwarfed size of individuals as compared with their representatives in the Mediterranean or Atlantic.

To these forms must be added *Dreissena polymorpha*, which has now established itself in most of the rivers of Western Europe, but of which the native home is in this region, and the dreaded *Teredo navalis*. The abundance of this Mollusk in the harbour of Sebastopol is so great, and the destruction of the vessels which it attacks so rapid (eight years being the average duration of the under timbers of any ship), that it is possible that a great service was rendered to the naval power of Russia when it was compelled to withdraw its fleet from the Black Sea waters.

Some peculiar forms of *Cardium* occur in parts of the Black Sea, and which are common to the Caspian. *Cardium plicatum* is found at the mouth of the Dnjestr, and *C. coloratum* at that of the Dnjepr and the Don.

The deficiencies in the Black Sea fauna are remarkable. All those classes of Mollusca which, as we have seen, are but poorly represented in the Eastern Mediterranean as compared with the Western, are here either altogether wanting, or are of rarest occurrence, such as Cephalopods, Pteropods, and Nudibranchs. Echinoderms and Zoophytes are absent. The composition of the water is inimical to all these forms. The Medusæ are represented by shoals of the common gregarious *Aurelia*.

The fishes of the Black Sea are very indicative

of this estuarine character of its waters. As compared with those of the Mediterranean, the number of specific forms is remarkably small, whilst that of individuals is marvellously great. Pallas notices the "Red Mullet" and the "Cuckoo Gurnard." The beautiful *Umbrina* of the Lusitanian coasts (*U. vulgaris*), and which is recorded as having been once or twice taken on our coasts, is among the rarer fishes of this sea. Several species of *Sparus*, with Blennies and Wrasses, are such as have been noticed in the Ægean. The Grey Mullet, the "Kephalos" of the Greek fishermen, is the common fish of the markets of Constantinople. It is met with in great shoals along the whole coast of the Black Sea, from Kertch to the Bosphorus. These shoals are composed of fishes of the same size or age. The little Atherine, which, as we have seen, is abundant in the Ægean, migrates into the Black Sea in the spring. Having passed the straits of Constantinople, the shoals turn northwards, keeping close in to avoid the current which sweeps down towards the outlet. Should there be an onshore wind as they pass along, which not unfrequently happens, enormous numbers of this fish are thrown upon the coast and perish.

The Gar-pike is common, as are Dabs and Flounders; these last also occur plentifully in the Sea of Azof.

The migratory and gregarious Tunnies (this general designation includes several species) pass upwards from the Mediterranean in the spring. The value and importance of this fish to the Byzantines (for it is the Tunny which fills their "Golden Horn" to overflowing) have caused its habits to be closely observed from early times; from these notices we find that its route is the same now as then, and that it still continues to fill the bay of Constantinople with its countless shoals with the same periodic regularity as it did 2000 years ago. The old Mediterranean Greeks thought that Byzantium was the home of the Tunnies; the present race of fishers know much better. This annual passage into the Black Sea and back again, is only the last stage of that long migration which the Tunnies have to perform. They are all Atlantic fishes, and rather Lusitanian than Celtic, though some few reach our coasts. They make their appearance about the Straits of Gibraltar and in the Western Mediterranean in the early spring, and travel steadily eastwards. From the circumstance that the fish taken about the islands of Corsica and Sardinia are remarkable for their size as compared with those which compose the shoals which follow the shores of Europe on one hand and those of Africa on the other, it is a part of the popular belief respecting the Tunnies that they move along the Mediterranean in three columns, of which the middle one consists of the oldest and strongest fishes.

The passage of these shoals along the coasts of southern Europe is a busy time, and one of general excitement, with the fishing population; the Pilchard fishery on our Cornish coasts is something like it, in a quiet sort of way. The Tunny is taken about Sicily and the Mediterranean generally the whole summer through, but then these are usually full-sized fishes, and it is very probable that the duty of performing the Black Sea pilgrimage is felt up to a certain time of life only. In the autumn the shoals return again. The fishermen maintain that these shoals are composed of fishes of the same year: the uniformity of size is certainly very striking. Still more, they profess to know the shoals as they pass back again, and can tell how much the fish have gained in weight in the course of the summer. As with birds, so with fishes—some migrate locally, some to remote regions. The distance from the Straits of Gibraltar to the Sea of Azof is not less than 2800 miles. Such is the migration of the Tunnies. Man looks out for them at every point along their course as they go; and as they return they are the food of countless thousands of the Mediterranean populations. As they pass into the Black Sea the Dolphins and predaceous fish which have followed them along their whole course, still pursue them, flocks of sea-birds hover over them; yet the living stream flows on, age after age, and seemingly with undiminished fulness.

The Sword Fish is taken in great numbers in the bay of Constantinople.

Most of the fishes which have been here enumerated, such as the Grey Mullets, the Gobies, the Atherines, and flat-fishes, are well known to us as being found elsewhere, in estuaries and brackish waters; but it is by the next series, or by the relative proportion of the cartilaginous fishes, that the Black Sea and Sea of Azof are mainly characterized.

Though the Sturgeon (*Aciperser sturio*) is taken in all the Atlantic seas of Europe, yet it nowhere can be said to be a common fish. It is only a rare visitant to our coast, and the specimens taken are always adults. It becomes somewhat more frequent in the marine province to the north of ours, and also in the Lusitanian and Western Mediterranean region, but it is by no means frequent there. The habits of the several species of this genus are not the same. The common Sturgeon is the most pelagic, or the greatest wanderer; but, considering the great distance at which they are taken up the courses of the rivers which empty into the Black Sea and Caspian, that they spawn in these rivers, and hibernate there, they would seem rather to be river fishes which descend periodically to the sea than sea fishes which ascend rivers. Though they are captured more frequently in the larger Mediterranean rivers than in Atlantic ones, yet even there they only occur as single fish.

As many as five species of Sturgeon have been distinguished, and they all belong to that great system of rivers which flow south and east into the Caspian and Black Seas. They are nearly Russian as to their nationality, and would be quite so, but for the Danube. They are the forms of a geological region of great antiquity which has undergone great physical changes, the account of which will belong rather to the Geological History of the European area.

The Sting Ray (*Trigon pistanaca*), and another species which has not been determined, are found in considerable numbers in the Black and Caspian Seas, as also in the waters of the sea of Azof, which at times are nearly fresh. It may surprise some to find such fish inhabiting such a medium; but it will be found that this Ray commonly occurs throughout the Mediterranean on the mud deposit of the mouths of large rivers, and it may also be remembered that fresh-water forms of the genus occur in the great rivers of the South American continent.

Lastly, a species of Lamprey is taken in great quantities in the Sea of Azof.

CHAPTER VIII.

CASPIAN SEA.

THE Caspian and Aral are wholly inland seas, receiving the inflow of great rivers, but having no ultimate communication with any larger ocean : in this respect they resemble the Dead Sea, with this difference—that the waters of the latter show an excess of salt, whilst those of the Caspian are only brackish; the meaning of this difference will be explained presently.

The Caspian is European from the point where the great range of the Caucasus comes down to its coast, in lat. 40°, to the mouth of the Oural River, in lat. 47°. The area of this sea has been estimated at 140,000 square miles; but the region which bounds its northern half on either side still presents unmistakable evidences that its waters have at some time extended west as far as the mouths of the Danube, and eastwards to the Sea of Aral. This was not a continuous expanse of water, but rather a chain of lakes, of which the boundary lines and connecting links may still be traced in lines of cliff. Elevations near the mouth of the Volga, of rather more than eighty feet above the present mean level of the Caspian, are capped with rounded shingle and beds of sand filled with the peculiar shells of this sea. From these accumulations we may infer that the water over this area had a former level higher by a hundred feet at least than it has at present. The difference of level between the Black Sea and the Caspian has been put as high as about eighty feet, and as low as only forty; but whichever may be the correct measure, accumulation of water within the Aralo-Caspian depression of such an amount would again unite the seas, and that without the intervention of any local depression of the land—a course somewhat too often invoked by the geologist to explain such changes.

The Caspian, having no outlet, should present indications of a gradual increase in the depth and extent, in consequence of the vast volumes of water which annually flow into it. So far, however, from this being the case, its mean level is constant, and apparently has continued so for a considerable period, as the accession from all its tributary rivers is counterbalanced by the enormous evaporation of that region.

Evaporation alone is the agent engaged in reducing the level of certain internal seas below that of the adjacent ocean. But for its communication with the Atlantic the Mediterranean could not maintain its level, and this consideration leads to an inference that the change which has taken place between the present time and that at which the Caspian had its former expansion has been a climatic one. At present the winter temperature of the northern Caspian falls much below freezing (10°–14° F.); even the south Caspian is colder than our English winters. In July the heat of Astrakan is equal to that of Sicily or the south of Spain. During three months the evaporation is very great, and the marginal shallow waters are warmed, whilst on the breaking up of the ice and the melting of the snow the waters rise, and are intensely cold. Such conditions are not very favourable for any large or varied Molluscous fauna, and accordingly we find that it consists mainly of forms which live embedded in mud.

This Caspian Molluscous fauna is as yet but imperfectly known; such, at least, is the impression which the assemblage of observed species produces. We miss the *Limnææ* and *Paludinæ* which may reasonably be looked for there.

Rissoa Caspia and two little *Paludinellæ* (*P. variabilis, P. stagnalis*) swarm in these waters, and the extent to which their shells must go to increase the sedimentary deposits of the bed of the Caspian is highly illustrative of the conditions under which such thick beds of Paludinella limestone were formed, as may be observed in the tertiary brackish-water formations of Maintz and other places. Associated with these are *Neritina litturata*, a *Mytilus*, and *Dreissena*, serving to complete the parallel.

The remaining Mollusks are all forms of *Cardium*. 1. *C. (Didacna) trigonoides*. 2. *C. (Didacna) Eichwaldi*. 3. *C. (Monodacna) Caspicum*. 4. *C. (Monodacna) pseudocardium*. 5. *C. edule*. 6. *C. rusticum*. 7. *C. (Adacna) læviusculum*. 8. *C. (Adacna) vitreum*. 9. *C. (Adacna) plicatum*. 10. *C. (Adacna) coloratum*.

1 and 2 are hardly distinct, and are most common throughout the whole of the Caspian. Nos. 3 and 4 are considered by Middendorf to be nearly allied, and Deshayes notices the resemblance of *C. pseudocardium* to the common Cockle (*C. edule*). Eichwald described the *C. Caspicum* from dead shells, and doubts whether this species is now living. The Caspian form of *C. edule* is small, but distinct. Dead shells of the variety *C. rusticum* occur in abundance, but it is supposed that this form may also have recently died out there.

The following are rather south Caspian shells: *C. vitreum* and *C. læviusculum*, which latter is thrown up after storms near Baku, in such quantities as to serve as food for pigs, cormorants, and other water birds. *C. edentulum* is found in the north Caspian, but never living. *C. plicatum* occurs also in the Black Sea, at the mouth of the Dnjestr, but is there smaller than in the Caspian. *C. coloratum* is common to the Black and Azof Seas, and to the north Caspian.

The shells of the genus *Cardium* (Cockles), so numerous in all seas, as also at all past periods, are throughout remarkable for the constancy of certain characters. They are generally ribbed, the edges of the valves are crenulated and interlock; in the great majority of the species the valves shut close. The hinge consists of two central teeth in each valve, and two lateral, somewhat removed; in all, four teeth in each valve. The common Cockle (*Cardium edule*) is a good type of the genus.

The Cockles are mostly marine, but our common edible species is found in harbours and high up tidal rivers, where the water becomes brackish; in these cases the shells present several modifications: they are invariably reduced in size, are thin, and have their external characters less strongly marked. The Baltic Cockle (*C. Balticum*) presents such changes, as do also the Black Sea and the Caspian form of this species.

The shell known as *C. rusticum* (Chem.) is recognised by Philippi and Middendorf as a variety of the common Cockle—an aberrant variety, says Ed. Forbes, produced by the admixture of fresh water with the saline element. This variety is found in all European seas. In the Caspian the differences betwixt the *C. edule* and the *C. rusticum* are clearly marked only in the young shells; when older, they become so alike as to be scarcely distinguishable.

There is another aberrant form of *Cardium*, known as the Greenland Cockle, which lives in estuaries there, and although it no longer belongs to our European area, it is met with in abundance as a fossil shell in the crag deposit of Suffolk and

Norfolk, more particularly in the fluvio-marine portions. The peculiarity of the shell consists in its being thin and smooth; the hinge is nearly edentulous; rudiments of a single tooth in each valve may be detected in young shells which finally disappear. The animal is a true Cockle, but the shell is wanting in all the usual characteristics of the genus.

The Caspian Sea, with its very limited Molluscous fauna, makes us acquainted with another series of aberrant forms of *Cardium* in which the hinge undergoes great modifications, and which are accompanied by changes in the form and other characters, to such an extent that they have been referred even to other genera, such as *Corbula* and *Pholadomya*.

To these forms, but allied to them by corresponding modifications, may be added as many as twenty others, to which M. Deshayes has given distinct specific names, and which are found in the deposits of the older extensive Caspian. The hinge structure of this group taken altogether presents every conceivable deviation from the normal formula of the genus *Cardium*: the lateral teeth are suppressed, either one or both, and the central ones preserved; and the reverse take place. Often one tooth is alone preserved, and this is sometimes the anterior, and sometimes the other. This single tooth at times acquires a great development, and is accompanied by a great distortion of the shell on that side; indeed each change in the hinge structure has its attendant external change; and lastly, hinge teeth altogether disappear. The hinge structure as a generic guide altogether fails, and the shells take the external forms of *Isocordia, Venericardia, Crassatella,* and *Venus,* yet they are all true *Cardia,* and as aberrant forms are linked continuously one with another, and lead back to the *Cardium edule,* as the primary form.

Fishes abound in the Caspian. In no part of the world, Newfoundland excepted, are fisheries so productive, or do they give employment to so large a number of persons as they do about the mouths of the rivers which discharge into this sea.

Those principally taken are the great Silurus, the great and lesser Sturgeon (*Accipenser huso* and *pragmilus*), together with that most abundant but less-esteemed species the *Accipenser stellatus.*

Pallas mentions a circumstance which may serve to convey some idea of the vast numbers of fish which ascend the Sallian. The weirs for stopping the fish are established where the river is 160 yards broad and twenty-five feet deep. At these places as many as 15,000 Sturgeon are taken a day; but if the fishery is suspended for twenty-four hours, the fish so accumulate that they become packed, and fill the whole bed of the river to the level of its banks.

Lastly, Seals are as abundant in the Caspian as they are in the Black Sea; there are several species, but whether the common northern Seal (*Phoca vitulina*) ranges there is now doubted. These amphibians occur in great numbers over the whole of the lake system which stretches from the Black Sea across central Asia.

CHAPTER IX.

ON THE DISTRIBUTION OF MARINE ANIMALS.

THE assemblages of animals composing the fauna of the European seas have been shown to undergo a constant change from place to place. The differences and peculiarities are broad enough in certain cases to be seen by those who commonly bestow but little attention on such matters. Thus the peculiarity of the fishes of Mount's Bay would be recognised at once by any one who had only seen those captured on the coasts of Norfolk; and the abundance of the elegant Nephrops in the Dublin market would arrest the attention of the epicure whose experience had been limited to the edible Crustaceans of the London fish-stalls.

This system of change, or of geographical distribution, has of late acquired a great amount of interest; it is connected with some of those curious inquiries into bygone conditions of the earth's surface which are now undergoing investigation, so that the subject forms a necessary part of the natural history of each separate province; and although we may not as yet have arrived at a satisfactory knowledge of all the conditions which bear upon this enquiry, it may not be amiss to consider some of those influences which obviously regulate, in some degree, the distribution of marine life, and the changes which they immediately produce.

Foremost amongst these is the influence of temperature. The marine fauna which we have been here considering occurs on a line of coast which, if limited to European countries, has an extension in latitude of nearly 3000 miles, but which zoologically extends from the Arctic basin to the Canaries. It will be sufficient in this place to notice the winter and summer temperatures of successive sections of the Atlantic coast-line, and to connect these with the condition of the internal seas. On the Russian shores of the Arctic Ocean the mean cold for the two winter months falls below 5° Fahr.* This is the winter temperature of Spitzbergen, and the coast is ice-bound from October till May; yet here, as we have seen, and at depths below the reach of ice, there is a Molluscous fauna.

Compared with this, the temperature of the west coast of Scandinavia exhibits a great change, and is comparatively mild; from Cape North, nearly as low as Bergen, the degrees of cold range from 23° F. to 32° (freezing), but, at which place, the sea water does not freeze oftener than three times in a century. The portion of the coast where the lower temperature prevails, from Cape North to the Lofoden Islands is that along which the characteristic fauna of the Arctic basin reaches.

* See on map the course of the blue lines.

Iceland, which has the winter temperature of North Cape, has also its Arctic assemblage of Mollusca.

The Baltic area experiences a degree of winter cold far below that of the portion of the external coast, corresponding to it in latitude. At the upper end of the Gulf of Bothnia the temperature is that of the Arctic coast; in the Gulf of Finland there is the cold of Cape North. Such are the low temperatures affecting the brackish waters of the northern portion of this internal sea, and which may in part account for the poverty of its fauna as compared with that of the southernmost half.

Crossing the whole European area, the great internal brackish seas of the Aral and of the north Caspian, have winter temperatures corresponding with those of the Arctic Ocean.

It is from about Bergen and the southern parts of Norway and Sweden that the assemblage of Mollusca and other marine animals had been obtained which form what has been termed the "Boreal" or "Scandinavian fauna." This section of the western coast of Europe, with a somewhat higher winter temperature, shows a wonderful increase in the numbers of the component members of its fauna.

The whole of the group of the British Islands, our internal seas, and both coasts of the English Channel, come within a winter temperature of from 40° F. to 32° F.; such is the winter cold of the higher end of the Adriatic, as also of the southern portions of the Black and Caspian seas.

The Celtic province of Ed. Forbes was formed to include under one term a fauna which, under many favourable conditions, becomes peculiarly rich; southern forms begin to show themselves, and nowhere is the direct relation of distribution to temperature better shown than here.

Pursuing for the present the subject of low temperatures, sea water, as is well known, seldom freezes in our Celtic region; when this happens, as it did in the winter of 1854–5, we had a good illustration of the effects of cold on a portion of a marine fauna. The shallow pools of water over the interval between tides and the surfaces of the mud-banks with their growths of weed, were all frozen hard; the animals frequenting this zone mostly perished, and for months afterwards there were parts of our southern coast where lines of littoral shells, with their putrid contents, stretched in thick bands along the upper tidal line.

A writer in the "Witness" newspaper, perhaps the late Hugh Miller, gave a graphic description of the effects of the cold of the same winter on the Mollusca of the Frith of Forth. Oyster farmers know full well, to their cost, the havoc of a few hours' cold on their uncovered fields.

The weedy surfaces of our mud-banks swarm with small molluscous vegetarians, whole tribes of which have their limits there. When, as on our western coasts, or on those of France or the Channel Islands, hard rocks face the sea, what rich gathering grounds does the sea-side naturalist there meet with, over miles of broad, horizontal ledges! A few successive winters, with low temperatures, would destroy the whole of the fauna of this broad intertidal zone, would alter the present relative proportions of littoral species, and give another aspect to our marine testacea, viewed collectively.

A mean winter temperature of 54° F. includes the north, west, and east coasts of Spain, together with Sicily and Greece. Gibraltar, and the southern shores of the Mediterranean, are warmer by several degrees. From the Arctic Ocean to the mid-Mediterranean there is a difference of mean winter temperature of more than 50° F.

The influence of summer temperature is best indicated by the range of southern forms. That part of our own coast which just comes within the July mean temperature of 64° F., namely, the extreme south-west parts of Devon and Cornwall, is also that from which our rarer southern forms of fishes and Mollusca are taken. The researches of Mr. M'Andrew have shown that Vigo Bay has an isolated assemblage of testacea of a somewhat northern character; it may be an outlier of the marine fauna of a former period, but its present distinctness may have been maintained by the somewhat lower summer temperature of Gallicia, compared with the rest of Spain and the Bay of Biscay. In like manner it is not supposed that the southern forms of the south-west of Ireland have migrated there from the south, across the deep waters of the opening of the channel, but that their presence there, so far as present influences are concerned, is dependent on the peculiar local conditions of that coast as to temperature.

These considerations lead to the inquiry as to the meaning of those local assemblages which have been observed in several parts of our Celtic province, the existence of which was first detected by Ed. Forbes, and for which he proposed the name of "outliers."

" At certain spots we find assemblages of northern forms, so peculiar and so isolated, that we cannot account for them by any facts connected with the present disposition of currents, or other transporting influence." These patches are especially to be met with in the Clyde district, and among the Hebrides; on the east coast in the Murray Frith. It is probable there is another patch somewhere near the Nymph Bank, on the S.E. coast of Ireland, and another in the German Ocean.

These "outliers" are usually located in a hole or valley of considerable depth, from eighty to beyond 100 fathoms, and consist of assemblages of Mollusks, of more northern character than the zone or province in which they occur. The species which Ed. Forbes cites, are *Cemoria Noachina*,

Trichotropis borealis, Natica Grœnlandica, Astarte elliptica, Nucula pygmœa, Terebratula caput serpentis, Crania Norvegica, Emarginula crassa, Lottia fulva, Pecten danicus, Neœra cuspidata, N. costata, and *N. abbreviata,* being such as are met with together in the far north (pp. 49–58).

The explanation which Ed. Forbes gives of these "outliers" is as follows:—When the bed of the sea of that period when in our latitudes the fauna was more northern than it is now was upheaved, the whole was not raised into dry land, but tracts of greater depth, and which consequently were tenanted by peculiar forms, still remained under water, though under different depths. In these changes a portion of a fauna would be destroyed, but such species as could endure alterations in vertical range would live on.

If the following diagram, A, represents the relation of sea to land for the period of the northern fauna,

A

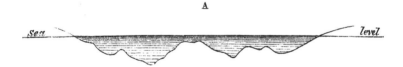

the next, B, may represent it after the partial upheaval of the sea-bed. In this last, the unshaded interval below the water-line will be that in which the newer fauna has established itself in shallower waters, and the shaded part that in which the remnants of the northern are supposed to be isolated beneath the newer and existing fauna.

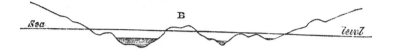

If such be the real meaning of these local assemblages of northern forms from depths about our islands, there will necessarily occur areas of sea-bed at more moderate depths, where residual portions of the northern are associated with the present fauna. One such has been noticed by Mr. Jeffreys. The Turbot bank off the coast of Antrim, in twenty-five fathoms water, gave twenty-one species of Testacea, "Arctic," "Boreal," "Celtic," "Lusitanian;" all there assembled together.

Zoological "outliers," therefore, can only be looked for where the existing marine fauna is a compound one,—the result of the admixture of forms from adjacent areas; they imply changes of conditions over the areas in which they occur, both as regards temperature and depth; and inasmuch as there is a tendency to uniformity at great depths, the differences between provinces being mostly found in the sublittoral zone, it follows that, though there may be outlying southern species in northern provinces, yet there can only be distinct northern assemblages of species beneath seas which, in the progress of change, have become warmer. A very great amount of change in latitude is necessary before a complete change is brought about in the species inhabiting deep-sea zones; the interval between the Arctic circle and the Tropic of Cancer does effect it. Residual deep-sea forms of tropical assemblages cannot, therefore, be expected beneath such as belong to the higher sea-zones of more northern assemblages. The foregoing considerations may be of use to the palæontologist and geologist, and will frequently be referred to in the sequel.

Isolated groups of fossil remains are not uncommon amidst our old sedimentary beds; a remarkable instance has been noticed by M. Barrande. In one of the "lower divisions" of the great palæozoic series of Bohemia, he has described the occurrence of a patch of as many as sixty species, which forms do not agree with those characterizing the "lower division." These forms have lived in the beds in which their remains are found; they ultimately cease, and have been surmounted by beds which contain the forms of the "lower division." These sixty species are isolated, but they appear again as a component part of the fauna of the "upper division" of the same palæozoic series.

To these isolated assemblages of upper palæozoic amidst lower palæozoic forms, M. Barrande has given the name of "colonies." They are true "outliers," and will serve to suggest curious and interesting geological inferences in the earlier history (both natural and physical) of the European area.

The tendency of a body of water is to keep its surface temperature in equilibrium with that of the air which rests immediately on it. But numerous observations have established that the mean temperature of the surface of the ocean, from the equator to about 50° of north and south latitude, is somewhat warmer than that of the air.

There is a line extending from one Polar region of the earth to the other, at which an invariable temperature of 39° F. is met with; the depth of this temperature from the surface varies with the latitude; at the equator it is at a depth of 7200 feet, and it rises to the surface in lat. 66°, N. and S.

It has been seen to what an extent the richness of the Atlantic fauna is increased in a direction from N. to S.; this increase in the variety of specific forms, which so characterizes southern latitudes, takes place in the marginal and submarginal zones, and may be considered to be immediately dependent on temperature.

The line of uniform temperature sinks from the surface towards the equator at the rate of about 130 feet for every degree of latitude, so that, apart from the conditions of light and pressure, there is a definite point in every latitude at which Arctic and Boreal forms meet with their congenial temperatures; and hence a strong *à priori* probability of geographical distribution of Arctic forms, according to bathymetrical lines of temperature. An animal requiring for its existence a temperature of 29° would have to sink upwards of 100 feet for every degree it migrated south. In like manner, should a change in the temperature of any marine province be brought about (and this has happened repeatedly over the European area), as, for instance, from cold to warm, the marginal forms could only continue their existence by moving to greater depths. The case is here put in a purely hypothetical form, for no marine animals are so exacting in their requirements; very many forms have a range in depth in the same latitude, which is considerable, and many have a broad horizontal range. Such being the real condition of the question, a general, and not a close relation of distribution in depth to distribution in latitude is all that can be expected.

Mr. M'Andrew has made an interesting observation, illustrative of the distribution of species, from temperature dependent on depth. At Mogadore, on the west coast of Africa, in lat. 31° 30', he obtained 110 species of Testacea: of these about one half range north as far as to our British coasts; when, however, the 110 species were divided into two sets, according to depth, eighty-eight ranged from the coast-line down to depths of upwards of thirty fathoms; amongst these are all those species which are characteristically African or Lusitanian. Of twenty-two species dredged in thirty-five to fifty fathoms water, all but six were well-known British shells.

COMPOSITION OF SEA WATER.

Specimens of sea water from the open parts of the Atlantic are very uniform in their composition, whether taken in the latitude of Gibraltar or of the Hebrides; but such is not the case with the waters of internal seas, nor again between the waters of the coast and those of the offing. The sublittoral sea-zone is that of the maximum of marine life, and it is along the coast-line that all those changes are to be observed, from super-saline, normal, brackish to fresh, which are severally dependent on the amount of surface evaporation, the influx of rivers, and on the equalizing action of winds and tides.

The density of water taken at the surface is less than of that taken at depths; the degree of saltness, also, increases in the same direction. The water from the surface contains less air than does that from depths, and the difference may equal one hundredth of the volume of water. Again, analyses of the waters of the Black Sea, the Sea of Azof, and the Caspian have shown that, though the salts which they contain are the same, the proportions are different. These varying conditions have a marked influence in local assemblages of marine animals.

From a series of observations taken within depths of eight fathoms, Admiral Smythe puts the temperature of the Mediterranean surface waters at rather more than three degrees higher than those of the Atlantic for the same latitudes. This condition would affect the composition of the water. To what extent it does so, whether the degree of saltness of the Mediterranean waters is greater than that of the Atlantic, is yet an unsettled question. There is no doubt but that there is a difference between the waters of parts of the Eastern Mediterranean and the Western, owing to the influx from the Black Sea, sufficient, as has been shown, to produce a marked influence in the fauna; but the latest authorities find no difference between the degree of saltness of the Western Mediterranean and of the external Atlantic Ocean.

Along the outline of all our seas, wherever there are deep indents into the land into which rivers discharge, or where the set of tides or other causes have run out banks of sand and shingle in advance of shelving coast-lines, the composition of the included waters undergoes variable amounts of change: in some regions brine-lakes are produced; in others brackish-water estuaries and lagoons. These last are the favourite resort of the keener sportsmen of all countries; fishes abound, as do water-birds, from the land side and the sea. They have their peculiar testaceans, whilst purely-marine species pass in and exhibit the power which certain forms possess of adapting themselves to altered conditions, to how great an extent they can change their habits, and what curious modifications their external forms can experience.

An acquaintance with these intermediate areas, and the zoological features they present, is indispensable to the physical geologist. Throughout the long series of old secondary and tertiary formations, like conditions are constantly presented.

Apart from minor areas of brackish water the North Atlantic passes, in the ultimate portions of two of its great lateral branches, into internal seas, which differ from estuaries only in their extent. They communicate with the ocean, and are brackish from the fresh waters poured into them.

The brackish-water fauna of our European area varies much according to the province to which it belongs. In our Celtic region *Rissoa, Assiminia, Neretina, Conovulus,* and *Truncatella,* habitually prefer the mouths of estuarine rivers, and low, seaside pools. *Littorina littorea* ranges away upwards from the pure sea water, but seems to suffer from the change. *Limnœus pereger* ventures downwards. *Scrobicularia* and *Mactra solida* may be taken as good characteristic estuary shells; they are at their maximum in such places, and attain their largest dimensions. *Cardium edule* is common, as is *Mya arenaica*, but both dwarfed.

"When visiting," says Ed. Forbes, "the great South Arran, in company with Mr. Thompson, we found an interesting variety of the *Cardium edule*, in a brackish lake, at the northern end of the island. The shells were remarkably thin and brittle; the animals were not buried in the sand, but inhabited the *Conferva crassa*, in which the majority of the specimens were found creeping about."

In warmer regions, as in the South Lusitanian province, our brackish Gasteropods are replaced by species of *Cerithia*, *Melania*, and *Ampullaria*. *Corbula*, though a deep-sea form at times, follows the habit of *Mya*, and in warmer regions makes its appearance in brackish waters. A gradation of form may be traced from typical *Corbulæ* to *Potomomya*, according to the medium in which the forms occur. This is an interesting fact, because true *Corbulæ* will be found at the interchange between salt and fresh water conditions as far back as the commencement of the cretaceous series, and again, in the fluvio-marine beds of the tertiary series of the Isle of Wight, where *Potomomya* also occurs in most wonderful profusion.

That like conditions produce like assemblages may be seen by comparing the Black Sea fauna (p. 201) with that of the Baltic (p. 88).

In the Gulf of Bothnia many of our common English air-breathing pond-snails have habituated themselves to the slightly-saline waters of that part of the Baltic; such is also the case in the Sea of Azof. The changes produced by the degrees of saltness of the water on certain fresh-water forms have been noticed by Ed. Forbes in his observations on the coast of Asia Minor. In this region there have been repeated interchanges of fresh, brackish, and salt waters, and the results are shown in a remarkable manner in the genera *Paludina, Melanopsis*, and *Neretina*. These genera present three series of peculiar forms, so different, " that at first examination we appear to have before us very distinct and well-marked species." He was satisfied, however, " that they were the same species, assuming Protean variations," under varying conditions of the medium in which they lived.

The nature of a coast-line, and the composition of the deposits which form the sea-bed at different zones of depth, are conditions which exercise an important influence on the general character and abundance of marine life.

The Testacea that live attached, or which perforate cavities for themselves, require hard strata: on rocky and stony coasts *Mytilus, Chiton, Patella, Haliotis, Cypræa*, and others are found. Some boring shells require or prefer limestone rocks, such as *Gastrochæna, Saxicava rugosa*, and *Lithodomus*. Others, like the *Pholades*, are as often found in pure sand-stones.

The amount of weed in the upper sea-zone determines the numbers of the *Phasianellæ, Rissoæ, Lacunæ*, and *Littorinæ* which a fauna will have. Granitic coasts, or those of hard slates or sandstones, seem to afford attachment for a greater quantity of marine vegetation than do limestone rocks. Sands favour the genera *Panopœa, Mya,* the *Solen* tribe, *Donax, Tellina, Mactra, Tapes, Venus,* &c. This, however, depends in a great measure on the description of sand. A very large proportion of the bivalved Testacea, of all seas, occur over a sea-bed of muddy sand; but there is a zone of clean sand in advance of most lines of coast which comes within the range of the tidal and wave disturbance of the water, where deposits are being formed, which, after a while, are broken up again, and which may be called the drift-sand zone. This is wholly unfitted for marine life, and the only organic forms it ever contains consist in the fragmentary shells and tests of other zones. I have dredged along a band of this kind for thirty miles on our own coast without finding a single living form.

In muddy lagoons *Scrobicularia* is abundant. *Neæra* and *Isocardia* prefer deep-sea mud.

On our coasts lines of shingle, passing down into running sands, are not prolific in animal life. More in the offing, and in situations where scollop banks have established themselves, there is usually found a rich and varied harvest of Mollusks and *Ophiuræ*.

Rocks rising somewhat abruptly out of deep water cannot be dredged, nor indeed can a rocky sea-bed, but the multitude of dead shells met with in the vicinity of such submarine conditions shows how favourable they are for supporting life. Gasteropods abound in such situations, and I have known the dredge to come up with little else than the fragments of branching *Bryozoæ*.

The species of bivalved Testacea have a wider distribution than the Gasteropods, but the relative proportions of these two great divisions depend, in every local fauna, on the nature of the coast. It is owing to this cause, according to M. D'Orbigny, that the inequality is so great in the shells of the Canaries. These islands are rocky; hence the number of creeping Gasteropods, whilst of the bivalves a large proportion consists of such as attach themselves,—*Ostrea, Chama, Spondylus*, &c.

Both sides of the North Sea, from the Murray Frith to the fiords of Southern Norway, if at any time they should be raised, with their sedimentary deposits, into dry land, would be found, though more than 300 miles apart, to contain an assemblage of marine Testacea specifically identical.

Over and along the coasts which encircle the Arctic basin, there is also for the northern shores of the Old World and the New a perfect identity of specific forms; and the same Arctic forms are common to the west coasts of Finmark and the northeast of Greenland.

The great Mediterranean fauna is distributed with wonderful uniformity, as is also that of the Red Sea.

Such areas, in respect of the identity of the species they contain, may be termed Isozoic.

The northern coasts of Massachusetts have Testacea, of which one-half are common to our European side of the Atlantic, and which belong to our "Boreal" province. These two opposite sections are only isozoic in degree, but they are equivalent, and may be called Omoiozoic.

As northern forms decrease in number from north to south along both sides of the Atlantic, the proportion of common species decreases; still a correspondence is maintained by representative forms rather than by identical ones, and the system of omoiozoic zones is continued even when, as in the case of the Canaries and the Antilles, there should be only two species in common.

The application of such considerations as these by the palæontologist to his own special inquiries is easy and interesting. There was no greater amount of uniformity in past times than there is at present; distribution has ever been influenced by the same laws. If on investigating old sea-beds it shall be seen that there are areas over which the fauna is uniform or isozoic, whilst in other directions it presents change, we shall be justified in seeking the explanation in the causes which produce like results at present. If, for instance, the localities of the great Saurians (*Enaliosaurs*)—the monsters of the secondary seas—are found to be northern and temperate, but not southern, we may be allowed to infer that the distribution of these forms was somewhat that of our existing Cetaceans, and that they belonged to what may be termed the Boreal and Celtic zones of the oolitic seas.

In like manner, taking lower forms, the northern range of *Nerinœa*, and other Mollusca, can be indicated for the oolitic and cretaceous seas, as closely as can that of *Cymba* or *Solarium* in our own European seas now.

The great Indian Ocean contains an assemblage of forms under every class, which gives to its fauna a distinctive facies; so, also, does the North Pacific: the North Atlantic fauna is distinct from either. These great divisions of the ocean admit of minor ones, or, as they have been here called, "Provinces." Those proposed by Ed. Forbes for the Atlantic coasts of Europe are the Arctic, Boreal, Celtic, and Lusitanian. It must not be supposed that in forming these he limited them by definite lines and boundaries, for no such hard lines exist. Change is throughout progressive; but, when sections of the European coasts are taken at wide intervals—if, for instance, the fauna of our Channel Islands is compared with that of the Lofoden group—the distinctiveness is seen to be very great. Whether the intermediate section of the Atlantic, between these two localities, forms one province or more, is a question which every naturalist will determine for himself, according to the amount and kind of distinctiveness which, in his opinion, a province should have. Such divisions, at best, are merely conventional ones, and the degrees in which provinces will differ, will depend on whether their number be large or small.

Mr. Woodward considers that a province should have *one-half* of its species *peculiar* to it. If subjected to this test, our proposed European provinces are certainly too numerous; but though they may not be such as the rigid naturalist requires, and even may not present sufficiently broad characters to satisfy the general reader, they will still, in some respect, be found convenient. Strictly speaking, the Lusitanian and northern provinces alone comply with the rule as to proportion of *peculiar* species—so that the Celtic province, which is established on the mixed and intermediate character of its fauna, is not of like value with the others.

A province is distinct so far as it is supposed to contain a certain amount of specific forms which have not been found in some other part of the same sea or ocean, but it has never any stronger support than that of negative evidence.

The northern limit of the Lusitanian province seems to have been indicated by Ed. Forbes, when he states, "that the collector in search of a complete series of British shells, would have to go to the Channel Islands for those forms which, though included in our list, are almost extra-British." Since this was written the Lusitanian species, which have been ascertained to range as high as the prolific coasts of this group, have been somewhat increased, including the two magnificent Conch-shells *Triton nodiferus* and *T. cutaceus,* which, though common, are amongst the largest and most striking of Mediterranean forms.

Mr. M'Andrew has compared the results of his dredgings on the north coasts of Spain, including Vigo Bay, with those on the south. The British Testacea common to the north coast are 246 species in 406, or 61 per cent.; whilst the southern species are as 227 in 406, or 56 per cent.; and he further notices that, of the Scandinavian Testacea, which reach as low down as Spain, as many as 19 stop short, or do not pass south, of Cape St. Vincent. South of the same point, the character of the marine fauna becomes most obviously Lusitanian, so that, if it is thought desirable to reduce the number of independent provinces to two, it may, at the same time, be convenient to subdivide these; the Northern Lusitanian, in such a consideration as this, would extend from Cape St. Vincent to the Channel Islands.

When a marine fauna becomes specifically more numerous, as it always does (and always did) in a direction from cold to warmer temperatures, the rate of appearance and disappearance of forms in any direction is unequal. Of 212 species collected by Mr. M'Andrew on the north of Spain, only 29 did not extend to the south of Cape St. Vincent; out of 352 species obtained on the coasts of Portugal and

Spain to the south of that Cape, 140 species have not been met with so far north as Vigo ; if the tendency to diffusion was equal, the number here not passing north should be about 50,—or along the European Atlantic border the northern elements of the molluscous fauna have a greater southern distribution than the Lusitanian, or southern forms have northwards.

The naturalist who hopes that the day may come when some of the evidence as to the past conditions of the earth's surface may be interpreted, by the combined aid of the laws of geographical distribution, of the bathymetrical arrangement of marine animals, and of sedimentary matter, must make these, and all allied considerations, his special study.

A marine fauna is not a constant assemblage. In every latitude along the western shores of Europe, it has long been undergoing a slow rate of change ; southern forms have been extending themselves northward : the testaceous fauna of our western counties is far richer and warmer in its aspect than that indicated by those raised deposits, of comparatively recent origin, which fringe those coasts. Conversely, that wide extension of northern forms into southern latitudes which has been referred to, must not be taken as wholly referable to the present— antecedently to the present the tendency of northern forms was southerly, and some remain there now, the residual members of that migration.

A local testaceous fauna, as exhibited in a list of species from any part of the western coasts of Europe, is the product of repeated changes of distribution, which have taken place there, dating back into remote times.

The Celtic or British province, whether it be considered distinct, or as only a transitional one, bearing the same relation to the Scandinavian that the northern Lusitanian does to the southern, has a good physical boundary in the breadth of the English Channel. This division points clearly to the distribution of the component members of its fauna, to a commingling of species, by the extension of certain southern species northwards, mixing themselves with those of a northern character, which have enjoyed a longer and earlier tenure of the region. This, however, does not take place equally throughout the seas which surround the British Islands. Certain species range both upwards and downwards along our outward western coasts: they are the "Atlantic forms" of Ed. Forbes. These pass, in a limited degree only, into our internal seas—the Irish and English Channels, and German Ocean—just as certain West African forms do into the Mediterranean, and lead to the impression that these seas are of comparatively recent origin, and are as yet but partially colonized; in other words, that change is still in progress.

In like manner, Mr. Jeffreys observes, "My first

impression, on examining the Testacea of the Gulf of Genoa, was, that the fauna of the Mediterranean was mixed, and not peculiar to that sea. I found a large proportion of species which were familiar to me as British, and others having a more southern, and even tropical habitat. This led me to inquire whether the division into certain definite areas, which the late Professor Forbes distinguished by the names of Boreal, Celtic, Lusitanian, and Mediterranean, was well founded."

Testing these several divisions, or types, by the results of his own Mediterranean researches, Mr. Jeffreys states, that of the species supposed to be peculiarly "Boreal," he found several in the Gulf of Genoa, "such as *Chiton Hanleyi, Mangelia brachystoma* and *Neæra costellata*. Another (*Mangelia Leufroyi*, or *Boothii*) has been described and figured by Philippi as a recent Sicilian species, and a fifth, *Scissurella crispata*, I believe to be identical with the *S. decussata* of D'Orbigny.

"Of the second division, or 'Celtic' species," continues Mr. Jeffreys, "I met with *Tapes pallustra* (of which the *Venus geographica* of continental authors is a variety), *Acmæa virginea, Lucina borealis*, and *L. flexuosa*. Philippi has given *Trochus millegranus* and *Eulimella M'Andrei* (*Melania scillæ*), as Sicilian species. Of the third division, or 'peculiarly British' species, several, as *Jeffreysia diaphana*, and the so-called *Skeniæ*, besides *Argiope cistellula* (*Orthis Neapolitana*, Scac.), also occurred to me in the Mediterranean : and of the last division, or 'glacial' species, I detected three, *Nucula decussata*, *Neæra cuspidata*, and *Cardium suecicum*. Philippi has given *Arca raridentata* as Sicilian."

Mr. Jeffreys found that more than thirty species, which had been supposed to be restricted to our British seas, ranged into the Mediterranean.

The relation of the nature of the sea-bed and the associated Testacea to depth of water, was carefully observed by Ed. Forbes and the officers of the *Beacon*. For more than two months the dredge and the sounding-lead were actively employed for this purpose, in the Gulf of Macri, on the Lycian coast.

"Tracts of sand are forming near the shore, and off the mouths of the larger rivers. This is especially the case on exposed coasts, as in the instance of that part of the Lycian shore where the Xanthus empties itself into the sea. There the sea is shallow for some distance, and for a considerable breadth, the bottom being formed of a tract of sand. Such a bottom is not favourable to abundance or variety of marine life, and Testacea are by no means plentiful in such places.

"The muddy deposit from the deep sea is usually, almost invariably, of a pale yellow colour, and, when dried, nearly white. The region of this yellow mud is the sea-bed below eighty fathoms, more commonly below one hundred. From that depth down to as deep as we were able to explore by means of the dredge, we found an uniform bottom of fine sediment in the form of yellow mud, inhabited through great part by an uniform assemblage of marine animals, mostly delicate, fragile, and colourless forms, which became fewer and fewer both as to numbers and individuals, and number of species, as the sea became deeper and deeper."

"Beds accumulating around the bases of rocky submarine peaks, rising in deep water at a distance from land, are more likely to be embedded with organic remains, than such as are formed along shore. Round their bases will accumulate beds of shells and corals, belonging to various zones of depth. Such is the case, as we found by dredging, round the peak of rock in the neighbourhood of Cape Artemisium."

The distribution in depth of the molluscous and other forms, which were observed by Ed. Forbes in the Ægean, has been already noticed; that of the Testacea of our own seas is given for every species in the joint work of the same author and Mr. Hanley.*

Ed. Forbes thus subdivides the uppermost, or littoral zone. First, a line with the smaller varieties of *Littorina rudis* and *L. neritoides;* a second, with *Mytilus edulis* and larger forms of *L. rudis;* a third, with *Littorina littorea* and *Purpura lapillus;* the lowest, with *Littorina littoralis, Rissoa parva,* and *Trochus cinerarius,* accompanied on our west

* A History of British Mollusca.

and south-west shores with *T. umbilicatus* and *lineatus.*

To this zone, when rocky, belong *Patella vulgata, Skenia planorbis,* and *Kellia rubra.* In brackish water *Rissoa ulva* swarms.

The second is the "Laminarian zone," from the abundance of that and other sea-weeds, extending from low water to 15 fathoms. The genera *Lacuna* (one species excepted), *Calyptræa, Aplysia, Scrobicularia,* and *Donax,* do not range in our seas below this belt. *Rissoa, Chiton, Bulla, Trochus, Mactra, Venus,* and *Cardium,* are at their maximum here.

The third is the "Coralline zone," reaching from 15 to 50 fathoms. Vegetation is scarce, and ultimately disappears within these limits; and the zone takes its name from the hydroid zoophytes. *Trochus ziziphinus* and *T. tumidus, Chiton asellus, Acmæa virginea, Turritella communis, Venus ovata* and *V. fasciata, Pecten opercularis, Pectunculus glycimeris,* and *Nucula nucleus* mark the upper portion of this zone. *Solen pellucidus, Pecten varius, Modiola modiolus, Dentalium,* and *Mactra elliptica* occur lower.

Genera of Testacea have also their characteristic zones of depth. As we draw nearer to the present, in following out the sequence of fossil forms, a system of representation in time becomes distinctly marked, so that it would not be difficult to arrange the *species* of many tertiary groups of strata bathymetrically, according to the known conditions of existence of their present representatives. As we recede in time, the guidance of this *generic* distribution becomes safer and more available to the palæontologist.

The zone of the maximum of a genus is that in which it exhibits its greatest number of specific forms. In the Ægean, *Cardium* has its maximum between 20 and 35 fathoms deep, where it is represented by six species; *Pecten*, at between 60 and 80 fathoms, where it has eleven. In both cases, the zones in which these genera are most fully represented, numerically, are very different ones: all the species of *Cardium* put together do not amount to the individuals of the single *Cardium edule*, which occur within the first 12 feet from the margin; so, also, with *Pecten opercularis* at a somewhat greater depth.

The *Rissoæ*, as might be expected from their habits and food, have their maximum in the sub-littoral zone; there, also, they swarm numerically. Whenever these two conditions are combined, the palæontologist has a safe indication of marginal sea-bed. This genus has its deep-water representative, however. *Trochus* has its maximum in the Ægean in depths between 10 and 20 fathoms; but the excess in this case is very small, and the genus may be said to be fully represented from the marginal line down to 100 fathoms; generically, therefore, this genus is not very characteristic of definite depth. The *Pleurotomæ* have their maximum as low down as from 35 to 55 fathoms, and above and below this the specific forms decrease progressively: out of twenty-four species, one-half are dredged between those depths. None *live* in the marginal zone; and one (*Pleurotoma abyssicola*) was found below 100 fathoms.

In our own seas the *Pleurotomæ* belong mostly to deep-sea zones; so, also, over the intermediate region of the Lusitanian Atlantic; out of twenty-three species obtained by Ed. Forbes from the Ægean, one-third, at least, are British.

The study of the distribution of marine life according to zones of depth, suggests to the palæontologist many useful cautions; it teaches him that under different depths, and in the distinct deposits forming there, are assembled characteristic suites of animals, living apart, which when they die are entombed apart, and leave there the evidences of their past existence. These assemblages are as distinct from one another as are those which characterize the subdivisions of the deposits of older times, whether tertiary, secondary, or palæozoic.

The sublittoral zone of every sea and ocean presents the fulness of its fauna, and from that it decreases progressively and rapidly, till in regions far within those over which the finer sedimentary deposits are distributed, animal life altogether ceases. Far beyond the zones where the members of a marine fauna *live*, there are areas of wide extent, where animals of oceanic habits strew their delicate structures:

this is the zone of the "free-swimmers"—Pteropods, Nucleobranchs, Pelagic Cephalopods, and Crustaceans. The depositions of all past times present every gradation of bathymetrical distribution, down to the "Azoic zones" of depth, and the mere geologist must beware not to misinterpret the evidence presented to him, and suppose that some old world of waters was without life, merely because he finds no traces of it. Still less, on such negative evidence, must he speculate as to the "dawn of life," "Protozoic forms," and "Primordial zones." The history of our own seas, and of all seas, teaches us that there is a law of proportion in the classes and orders of the living things that dwell there, and that the presence of one form is safe ground of inference as to the co-existence of countless others. There may be no marine fauna older than that which the palæontologist has termed "Palæozoic," but it is most unphilosophical to suppose that organic life commenced with, and was limited to, *Lingulæ* in the latitude and longitude of Festiniog.

CHAPTER X.

EARLY HISTORY OF THE EUROPEAN SEAS.

It is still a favourite notion, constantly repeated under some form or other, that old faunas were comparatively poor; that as the world has aged life has multiplied; and that Nature now, in all her forms, whether of animals or of plants, is richer, fuller, and more varied than of old. The Natural History of the European seas affords no support for such belief. The fauna preserved in the palæozoic limestones and slates of south Devon, is infinitely more varied and numerous than that now found on our south-western coasts. The assemblage contained in the Crag deposits of Suffolk far exceeds the present fauna of the German Ocean. The Testacea of the Nummulitic period, such as may be met with from the south of England to the Mediterranean, probably exceed five-fold those which will characterize the present period for the same area. Of all the marine faunas which have succeeded one another in European latitudes, that of the present time is numerically the poorest. Under the same genera specific forms are fewer, whilst orders and classes which have been very fully represented in former periods, are very sparingly so now.

To such as are acquainted with the general outline of the earth's natural history, past and present —who know that our European tribes of animals and plants have not through all time had their being on our area—that before them have been numerous other assemblages in succession, and wholly different—many questions must have suggested themselves as to the manner in which the change from one fauna to another was brought about, how old forms disappeared, whence new forms came in.

Our knowledge of the remote past, with all its changes, is mostly derived from the accumulations of old seas, lakes, and estuaries; and if any satisfactory answers are ever to be given to the foregoing queries, they must be derived from a close and careful study of all the influences which determine the distribution and development of life now, within such-like areas. The inquiry is altogether distinct from those difficult questions, so often put—what is a species? and how do new species come into existence? and it resolves itself into this—where and what is that marine fauna in which we can first recognise the existing forms of our European seas?

Before entering on this inquiry, the meaning of the expression "a marine fauna" must be clearly defined; to say that it is such an assemblage as may be found living at one time in the same sea is not sufficient. Many bygone assemblages of marine animals—palæozoic, oolitic, cretaceous, and nummulitic—have in turn tenanted the waters of the Atlantic depression; and each has extended across the same zones in latitude as does our existing European fauna. If, with respect to the present, we limit a fauna to such forms as co-exist, no comparison with the past can be made; the two assemblages represent in one case a definite, in the other an indefinite, portion of time.

In those great assemblages known to the palæontologist as "the fauna of the Cretaceous period," or of the "Nummulitic period," are comprised forms of which we know that they did not all co-exist; and, further, that each period was marked in every latitude by the constant in-coming and outgoing of distinct species. We can ascertain the extent of many an extinct fauna as a whole, from its establishment to its close, though we may never know, except in a very limited degree, what were the relations of its component subdivisions.

That our existing European marine fauna may have a corresponding value with that of any of the great assemblages of the palæontologist, it must have a like extension; it must be dated back, so as to include all those forms which have co-existed since any species now found in the North Atlantic first made its appearance there.

Vast as are the periods of past time which the

phenomena of pure geology require and imply, they are brief comparatively with those during which a definite marine fauna has maintained its existence. It will be seen in the sequel how great are the physical changes which have taken place in the northern hemisphere, since a large proportion of our existing Testacea have occupied the Northern Atlantic.

The fauna of the European seas dates back its origin, or first appearance, to times which (on the scale of the geologist) follow next after the Nummulitic period (Eocene). So far as European seas are concerned, they do not contain a single species in common with the forms of the nummulitic group. The earliest records of the occupation of the Atlantic by any existing forms are certain old sea-beds, which are scattered at intervals over some of the western departments of France, extending inland along the valley of the Loire, as far eastward as beyond Blois, to be met with in some of its branches northwards—an old arm of the Atlantic, with dimensions nearly equal to those of our English Channel, long since laid dry. These old sea-beds are the "Faluns of Touraine."

Lower down to the south, from the Island of Oléron, across to the Adour, was another great indent of the Atlantic—an eastern extension of the Bay of Biscay. Over this once depressed area there are sea-beds which contain an assemblage like that of the Touraine deposits (*Faluns jaunes* of Grateloup).

The Testacea and Echinoderms from these two areas are somewhat peculiar. Extinct or unknown forms are in large proportion; these will be considered hereafter, but out of rather more than three hundred species of Testacea, there are about eighty which are identical with forms now living in the Atlantic.

These eighty species are not, however, now associated on any part of the French Atlantic coast; their localities are more southern. A better knowledge than we as yet possess of the Testacea of the West African coast would, in all probability, bring the Falun fauna and that of the present Atlantic into somewhat still closer relationship. For the present, the proportion of recent species may be taken at about 25 per cent.

Including existing forms, the facies of *the whole* of the Touraine assemblage is indicative of a more southern province than that which is now found in the parallel of the Loire (or 47° north latitude): estimated according to change in latitude, the difference may be put at from eight to ten degrees; or, in other words, at the time of the Falun Testacea, the warm zones of the Atlantic reached by so many degrees farther north than they do now.

The contrast between the fauna of the Atlantic coast of the department of the Loire and that older assemblage to be seen close by, is far greater than would be brought about by the shifting of the Lusitanian province northwards for its entire breadth.

The forms of the Faluns of Bordeaux and Touraine were not local or exceptional assemblages; they indicate directly that for a broad expanse in latitude the Atlantic, at that early stage, had a fauna of a more southern aspect than it has now, and they suggest further, that like general characters, modified by the ordinary rate of change, were maintained in its extension northwards. Although there are no broad areas which present remains of these older sea-beds, except at the entrance into the Channel (in the *Cotentin*), yet traces of the fauna, and of the period, are to be found in those outlying Lusitanian species which are to be met with about the Channel Islands, our own southern and southwestern coasts, or on those of the Atlantic border of Ireland.

It has been shown, by numerous illustrations, derived from various classes of marine animals, that the fauna of the Atlantic coasts of Europe is, for the most part, a complex assemblage; that from our own Celtic province, as low as the Canaries and the Mediterranean, it is composed of two distinct elements, a northern and a southern; and that the members of this middle group may be severally referred back to their original homes, whether north or south.

It is to be remarked, that the northern constituents of our present Atlantic fauna are not met with in the older fauna of the Faluns, nor in the equivalent assemblages further south. Northern forms had not, at that time, extended into that part of the Atlantic which lies west and south of the British Islands. Their great migration southwards took place subsequently to those great physical changes, which converted into dry land those portions of western France above referred to, and which changes were trifling in amount when compared with those of the same date in other parts of the Atlantic, and within the Mediterranean area.

The physical change which liberated the northern fauna has been indicated on independent considerations. It has been shown (p. 56) that there is good evidence of the former continuity of a coast-line from the north of Greenland to the north of Lapland, and that, consequently, the Atlantic did not then communicate with the Arctic basin; it was only when this barrier was removed that a free passage south was opened out to Arctic forms.

The breadth of this connecting link between the Old World and the New extended, probably, from 70° to 75° north latitude, and completed in its northern coast-line the symmetrical form of the Arctic basin. Sir John Richardson was the first to suggest both the existence and the date of this connection, in order to account for the remarkable agreement which the Boreal regions of the two continents present in their vertebrate fauna. In the small map which accompanies this volume, the extent of this subsided and submerged tract is indicated by dotted lines; but, as will be seen, this broad expanse of sea is marked by the emergence of land, at intervals, between the Western Islands of Scotland and the east coast of Greenland, at Iceland, and by the Orkney, Shetland, and Feroe Archipelagos. These islands, which have elevations of from 2000 to nearly 3000 feet above the sea, were the culminating points of this old terrestrial surface. Conclusive evidence of the continuity of land connecting these several groups of islands will be found in the common character of their flora, and in the relation of that flora to the Boreal and Alpine plants of the Old and New Worlds.

A full description of the botany of these North Atlantic groups, and of M. Martin's views respecting them, will be found in a former volume* of this series.

With the exception of this limitation at its northern extremity, the Atlantic is an old area of depression. There was an Atlantic ocean for the nummulitic, cretaceous, and palæozoic periods, during each of which it had its distinct zones of distribution in latitude, as well as its corresponding provinces of representative forms on its opposite sides. Our present inquiry is, however, relative solely to the growth or formation of that assem-

* Vegetation of Europe, by Arthur Henfrey, pp. 132-154.

blage of marine animals which constitutes our existing Atlantic fauna.

There are certain complex phenomena so immediately dependent on the physical arrangements of the earth's surface, that by assuming any definite changes in the conditions, it may safely be inferred what the results would be; thus the closing of the North Atlantic, in the quarter which has been indicated above, must have had precisely the same influence for that period that it would have now, should it be again closed.

So long as the Atlantic Ocean has had its existence, and reached from southern and equatorial regions as high as into 60° north latitude, so long must the equatorial current of heated waters have moved from east to west, have been deflected from the American coasts, and again made to cross the Atlantic. When the action of disturbing forces is now temporarily suspended, this current is found setting in upon some part of the western coast of Europe : such, however, is not its *ordinary* course. It will be seen, by reference to the map, that this broad ocean-river, our "gulf stream," after having flowed for a space of 50° from west to east, is suddenly turned due south in longitude 30° west, and becomes split up into minor currents and eddies.

This change in direction is due to the prevailing set of the Arctic currents. These are indicated in our map by arrows, which, it will be seen, point south-easterly from Davis's Straits, and more southerly for those setting out of the Arctic basin, through the interval between Greenland and Lapland. These Arctic currents thus converge towards the European shores of the Atlantic, and produce their effect just in proportion as their force is combined, and that of the gulf stream lessened by diminished velocity, as, also, by becoming expanded and shallower. It is only by the continuance of westerly and south-westerly winds that the warmer surface-waters of the gulf stream are occasionally carried forward and brought into contact with our western shores, bearing with them the vegetable products of the New World, together with the *Ianthinæ* and *Spirulæ* of the open Atlantic, as evidences of the course which the stream has taken.

When the Atlantic was closed at its northern extremity, there was no counteracting agency by which the stream of the equatorial waters could have been influenced, or their temperature reduced; and the constant flow of so large a volume of heated water sweeping round into this closed sea, must necessarily have imparted a great degree of warmth to the whole of the North Atlantic Ocean, giving a uniform and genial climate to its European border-conditions, which would materially influence the character of its fauna, whether terrestrial or marine.

Such, I imagine, were the precise conditions under which that early facies of our Atlantic marine fauna, which is to be seen in the Faluns of Bordeaux and Touraine, had its development. Like influences may be traced still further back in time, as into the Nummulitic period; in no other way than by the action of cross Atlantic currents can the western relations of certain forms found in both of these assemblages be satisfactorily accounted for.

The waters of the equatorial current raise the temperature of the central regions of the Northern Atlantic; and, from the prevalence of westerly and south-westerly winds, they thus, indirectly, influence the climate of north-western Europe; but, as a general rule, these heated waters do not now come into immediate contact with our shores—they are separated from them by a broad interval of sea, at a much lower temperature.

The removal of the land separating the Arctic basin from the North Atlantic, not only had the effect of lowering the temperature of the waters of the whole of that area, but, by the set of the oceanic currents which were forthwith established, the Arctic fauna became diffused along the whole of that Atlantic coast of Europe. The change was sufficient to extinguish—locally, at least—three-fourths of the previous fauna, as it had existed on the coasts of France and Spain, and it was at that time that the commingling of northern forms commenced, which has resulted in the present complex character of the marine fauna of our Mediterranean and mid-European regions.

If such was the character of the early Atlantic fauna of the Loire channel, and its favouring conditions, it may reasonably be asked whether any assemblages with like southern characters can be traced in other localities still further north, along our European coasts. Our own coasts offer a good example.

There are some old sea-beds high up our English Channel, near Selsey, on the Sussex coast, which, therefore, lie rather more than two hundred miles north of the Faluns of Touraine, and which, from geological position, are undoubtedly referable to a somewhat distant period. The Testacea of the Selsey and Touraine beds do not admit of strict comparison, for not only are the numbers very unequal, but the conditions indicated by the Sussex species are local and exceptional—such as muddy marginal lagoons contiguous to land. If these beds are of old date relatively to the present Atlantic or Channel fauna, it is quite sufficient for our present purpose if their contents show a deviation from the existing fauna, of the same kind as that indicated by the Touraine Testacea.

As yet we have only thirty-five species from the Selsey beds; of all these the relations are decidedly southern and western. Some forms, such as *Tapes pallustra*, all met with of large size, and another species, *Tapes aurea*, put on the aspect which they present at present in warmer Lusitanian and Mediterranean seas. But the two most remarkable shells of this deposit are *Lutraria rugosa* and *Pecten polymorphus*.

Both of these shells are well-known forms, and are exceedingly common throughout the south Lusitanian zone of the Atlantic, including the Mediterranean; but they have not as yet occurred further north than about Lisbon, which may be taken as their limit in that direction. Both of these species are good characteristics of the fauna of the south Lusitanian province, and in the early stage of the distribution of Atlantic fauna, they were fully as characteristic of the seas in the latitude of the English Channel, for the *Lutraria rugosa* was most abundant there. The distance which now separates these fossil and living forms of the same species is as much as four hundred miles, a distance as great, and in the same direction, as that in which the living representatives of the Touraine species have to be sought for.

Clear indications of the southern character of the early Atlantic marine fauna may be tracked still further north, and the temptation becomes great to dwell on the fossil contents of some of the old sea-beds of Ireland, which show so clearly, and for how long, that early fauna lingered on.

The representation here made of the early condition of the Atlantic, and consequent character of its fauna, is not a mere fanciful speculation, but will be found to be collaterally supported by many independent considerations.

Generic assemblages of plants and animals, whether terrestrial or aquatic, whether fresh-water or marine, have their regions, or definite geographical areas: these are what are known as "generic areas." Each of these has its "metropolis," or district of greatest number, either of tropical or specific forms; geographical unity seems to be one of the essentials of every generic group.

The genus *Mitra* offers a good illustration of this geographical grouping. These shells have their head-quarters in the Indo-Pacific Ocean; and they are thence distributed, but in decreasing numbers, in every direction away from that central region. Typical species of "Mitre-shells" from the Indian Ocean are to be met with throughout the Red Sea. Numerous other forms of the genus are found on the west coast of Africa, and about the Atlantic islands. As many as eleven species live in the Mediterranean, which are also mostly common to the Atlantic; but this is their present northern limit. These species do not range up the Lusitanian coasts, so that their European range is distinctly defined.

On going back to an earlier facies, or period of the present Atlantic fauna, "Mitre shells" of large and handsome forms are met with in the Faluns of Dax and Bordeaux, as in those of Touraine, where there are as many as seven species. Of these, one at least (*M. ebena*) is a well-known Mediterranean form; so that at that time the range of the genus was more northern than it is at present.

Mitræ occur on the Pacific coasts of America, but are seemingly wanting on those of its Atlantic border; but far to the north, in the seas of Greenland, there is a solitary form (*Mitra Groenlandica*), a seemingly exceptional case in the distribution of the genus. It is, however, just the kind of exception which serves to show the reality of generic centres.

This Greenland Mitra occurs fossil in Ireland, in association with another species of the genus (*M. cornea*) now living in the south Lusitanian zone, and it thus becomes linked with its congeners. It remains, there, to attest the furthest extension of its race, and is to the zoologist, when speculating on the former range of lower tribes of animals, just what the lonely Runic pillars, on the same Greenland coasts, are to the antiquarian, when engaged in tracing out the remote settlements of the early Northmen.

The palæontologist may derive much useful guidance from the study of generic areas. They will often enable him to determine the extent to which old seas may have been connected, whilst the occasional isolation of any definite forms, by intervals of deep and broad sea, is to him direct evidence of the former continuity of conditions, along which such forms have travelled—of physical changes of definite date, the best proofs he can have of the extent of these old seas, and of the modifications they have undergone.

Mitra Groenlandica is an eastern, or Old-World mollusk, in its generic relations, as are some others of the Greenland fauna; or as, in the case of the flora of the connecting land (p. 255), the long line of coast which once stretched from Greenland to Scandinavia, presented a fauna in which the assemblages of the opposite sides of the Atlantic were represented and blended.

Many considerations, however, make it probable that the species of *our* Boreal and Celtic provinces, which have western relations, exceed numerically those of the North-American Boreal region, which have had an eastern origin. This question will have to be treated somewhat in detail, when we shall describe those past conditions of the Atlantic, as well as those stages in its fauna which are indicated by the crag and other tertiary sea-beds. It may suffice, in the present place, to state that the somewhat indistinct features of the Arctic and Boreal faunas, if mere lists of species are taken, is the result of the migration outwards of the Arctic species. The Boreal fauna can be cleared of its mixed character, by separating all those species which at present have a wide range within the Arctic basin; and then the residual Boreal forms, which are not Arctic, and of which so many are common to both sides of the Atlantic, will represent the remaining portion of the fauna of that earlier stage of the Atlantic when it was closed at its northern extremity, so far as that fauna has been able to live on.

With a change so great as that here indicated, a large proportion of the original marine fauna of the Boreal province of either side of the Atlantic must have been totally extinguished, whilst of the forms that continued to live on, there are some that exhibit characters which deserve notice.

Pholas crispata is one of the shells of those Selsey beds of early date, which have already been referred to: as it occurs there, it is remarkable on account of its abundance, and great size, being more than as large again as any living specimens to be now met with in European seas. This shell, with like dimensions, is found fossil in Ireland, in some old sea-beds: as a species, it had its maximum at an early stage of the Atlantic fauna, and it has lived on. Ed. Forbes describes its present distribution, as a British species, to be Atlantic, ranging north into our Boreal province. It is one of the forms common to both sides of the Atlantic, and on the American coasts it occurs as far south as Carolina. This species therefore, though sufficiently common now in our European seas, may be considered to have had western relations originally, to have found more suitable conditions for its development in the earlier Atlantic, and more particularly in the portion north of latitude 50°.

Like our west European races of men, the members of the Celtic and Lusitanian marine provinces may be considered to have been derived from other regions; three-fourths of the whole assemblage may be thus directly accounted for, and this purely derivative character of the fauna belongs to zones which extend through more than thirty degrees of latitude. When southern and northern seas shall have been more diligently dredged, some others of the forms of this intermediate region will, doubtless, be discovered to be also immigrants.

The marine fauna of this broad zone of mixed races is as characteristic an assemblage, in the sense of the palæontologist, as that of the Arctic, or any other parent province; and it may be worth while to glance at the process by which a derivative fauna acquires a distinctive character.

There are certain Testacea, such as the common limpet (*Patella vulgata*), which seem to have their limits, or specific centres, within our European zone of mixed forms. This mollusk has its northern limit on the Norway coast, somewhere south of the Lofoden islands. It is not found within the Arctic province; it could not exist there now; and, unless the climatal conditions of that region should have been, at some time, greatly different to what they now are, it may be safely added that it never could have existed there. At present this form occurs in wonderful profusion in our Celtic province, more sparingly in the north Lusitanian, is scarce in the south Lusitanian province, is not found about the Atlantic islands, nor in the Mediterranean.

Tracing back the history of this species in time, it cannot be admitted into the fauna of the crag period, nor into that of the Faluns of Touraine— nor yet into any member of the Italian tertiary series. It makes its first appearance in the upper Faluns of Dax, under its commonest form, but by no means as a common shell. At this period the *Patella vulgata* must have been at its numerical minimum in European latitudes.

This form is at its numerical maximum now, as an element in the existing fauna of the European seas; it is referable to the temperate Atlantic (its specific centre is on the west coast of Ireland). It has a less southern range than it had before the communication with the Arctic basin was established; and it is since that physical change that it has found those conditions which have so favoured its numerical increase.

In all future time, *Patella vulgata*, in its profuse abundance, and numerous varieties, will characterize the deposits of a definite portion of the area of the existing European seas; and it will, moreover, have its definite place in the newer tertiary series. Many of our common species, when considered with reference to their range in time, and their distribution in space, will be found to have a history like that of the species which have been here taken as an illustration. The sum of these gives to a local fauna its character; sectional portions of a fauna acquire distinctive features, according to the number of specific forms which attain their numerical maximum, and development there.

The peculiarities of the marine fauna of the Channel Islands' group must have often puzzled the working naturalist: not only are things he there meets with wanting, in a great measure, on our own south-western coasts, but, so far as we know, they are also wanting for a broad band in latitude along the western coasts of France, and even to the south of Spain; so that if, as has been seen, the fauna of Vigo Bay is less Lusitanian than it ought to be from its position, that of the Channel Islands, on the other hand, is much more so; and the more we know of this local fauna, the more strongly does this peculiarity come out.*

The explanation of peculiar local assemblages has to be sought far back in time; and this is just one of those cases of geographical distribution in which it is necessary to call in the aid of geology. The area of this peculiar fauna is the coast of the Department of Finisterre (p. 90), forming the advancing foreland of France on the west, the coast

* See, particularly, "Gleanings in British Conchology," by M. Gwyn Jeffreys.—Annals of Nat. Hist., 1858-9.

of which, from the rocks of Porsal to the mouth of the Avranches river, runs due east for one hundred and sixty miles. This part of France is of old date in the earth's history. The chain of hills of the "Côtes du Nord," and those of the Boccage (Calvados), date back to times anterior to the "coal-growths," and have continued to form part of the earth's terrestrial surface ever since. The total absence of all secondary and tertiary deposits over any of the numerous islands which occupy the angle between the Cotentin and Finisterre, shows that it was originally part of the same raised area. The Channel Islands, and the numerous group of rocks, such as the "Roches Douvres," "Les Minquiers," and the Chausey islands, are the higher portions of this subsided land. It is to this area —which was not disturbed throughout the later half of the palæozoic period, nor through those of our oolitic, cretaceous, and nummulitic deposits, which formed the northern boundary for the old marine channel of the Loire valley, which was not affected by the changes which took place in the northern hemisphere during the later pliocene period—that we accordingly find that the earlier facies of the Atlantic fauna still belongs.

In the great Mediterranean basin we have the older sea-beds and their contents; taken by itself, part of this early fauna has disappeared or died out, and part survives either there or elsewhere;

upon this fauna new forms have come in, mainly of North Atlantic origin; the whole assemblage is seen changing its facies progressively from southern to northern, till coming down to more recent times the accession of immigrants is again West African. What is of interest here, is, that the relation of the existing fauna can be traced back stage by stage as it underwent change in the course of the tertiary period. The whole series, from the present back to the earlier fauna of the Turin beds, or of Montpellier, is the most complete and consecutive of any that we are yet acquainted with.

Palæontologists have taken marine faunas as the measures of the duration of geological divisions of time, and in this way the Mediterranean basin becomes the type of the true tertiary period. What the precise relations of the existing fauna may be to any earlier portion, are points which for the present can be considered as having received only a general answer; the tendency of recent investigation has been to increase the amount of agreement between the present and the earlier stages of the tertiary period. Mr. Jeffreys goes so far as to state that "it is most probable that every species which Philippi has described as inhabiting the coasts of lower Italy will eventually be discovered to have also had its existence in the tertiary epoch, and perhaps *vice versâ.*"

The disappearance of so many forms, and appearance of so many others, and that over so wide an area, are changes akin to those which the palæontologist traces amidst older sea-beds: commencing with the early history of our Atlantic marine fauna, we can follow the incoming and outgoing of a long succession of species, sufficiently distinct in themselves to admit of the recognition of the progress of change, yet connected throughout by such a number of common forms as to make the fauna indivisible as a whole. The duration of this fauna constitutes the true tertiary period of the earth's history, and the kind of change which it presents is precisely that, as will be seen, of every other great period, whether Cretaceous, Oolitic, or Palæozoic, when estimated, as geological periods ever have been, by the succession of local assemblages.

In the changes which our own European marine fauna presents, we are in some cases enabled to trace component parts to the localities or regions whence they came, or whither they have gone, and can, moreover, see the dependency of the zoological change on some definite physical disturbance. Our imperfect knowledge of the nature of the physical changes of the remote past does not as yet enable us to trace such a connection; as yet the palæontologist has hardly done more than note the local rate and order of zoological change; but all the considerations to be derived from the history of our European marine fauna tend to impress this, that, in all times, the nature of the process of local change may have been the same, and that it does not follow that forms have been newly created because they appear to us to make their first appearance at some given stage of a geological formation.

The single instance of the occurrence of a Lusitanian form, such as *Lutraria rugosa*, in our seas at an early stage of our fauna, its subsequent complete extinction in all seas, within a distance of four hundred miles in latitude, admits of useful application by the palæontologist. This, though a striking, is by no means a solitary instance. If the history of every component member of the marine fauna of our European seas was written in detail, from its earliest appearance downwards, they would all agree in respect of these apparent migratory movements in time, differing only in degree.

From the copious fauna which now tenants the Mediterranean waters, a series of changes may be traced, through older sea-beds of the same area, far back into bygone ages. Nowhere do we find better illustration than here of the nature of the change which a fauna may undergo in time : the evidence is consecutive. It is possible, however, that the Mediterranean series, recent and fossil, may be imperfect, and that the *earliest* periods of our European marine fauna are not represented there. A comparison of the contents of the older Italian deposits, and their equivalents, containing the remains of existing Atlantic species of Testacea, with those of the Faluns of Bordeaux and Touraine, suggest the probability that in these last we have an earlier stage still in the history of our fauna, referable to the time when the Mediterranean depression had not yet been opened to the Atlantic waters.

The precise relation of our existing fauna to any earlier stage, in respect of common species, is a large question, involving many subordinate points upon which naturalists and palæontologists have much to learn before they will be agreed; the tendency of more recent investigation makes the amount of agreement between the present and the past to be greater than was once supposed. Leaving aside the question, how new forms are introduced into a fauna, the history of our European seas teaches us thus much that is certain—that it is possible to point to contingencies under which the component members of a fauna may seem to migrate, disappear, and die out; that certain conditions of existence are so intimately connected with the continuance of the separate members of a fauna, that, unless these are maintained, their duration there becomes impossible. A small amount of change may cause such species to disappear, and, in all cases, the duration of any species over a given area will depend on its power of adapting itself to change. Hence the unequal terms of the existence of species in our present tertiary, as in all antecedent bygone faunas.

Another inquiry still suggests itself—To what

extent, it may be asked, have the Mediterranean and Red seas a community of specific forms?

This question has been partly answered by Philippi, to whom the Red Sea shells collected by Ehrenberg and Hemprich were referred for examination. He found that out of 382 shell-bearing Mollusks, as many as seventy-four were also common Mediterranean species. This appeared to be a somewhat large proportion, and, for the purpose of explaining it, some naturalists adopted the supposition of a communication between these two seas at some former period. This notion has had considerable weight given to it by some expressions of M. Deshayes in his notice of the Mollusca of the coasts of Greece. M. Deshayes, however, admits that the more the shells of the Mediterranean are studied, the closer becomes their connection with those of the Atlantic; so that it is rather by means of the fossil shells found in the older Mediterranean sea-beds, and which now form part of the lands of Greece and Italy, that this eastern connection is to be traced.

Certain species of Mollusks are so cosmopolitan that a low proportion of common forms may be expected even between two remote provinces, and will not constitute any difficulty as to the existence of Zoological provinces generally. A reference, however, to the list of species common to the Red Sea and the Mediterranean will suggest, both that the number may be somewhat modified, and also that

T

their present line of distribution may account for their presence in these two seas. The list is so short that it may be given at length.

* Solen vagina, L., Lus., Celt.
„ lugumen, L., Lus., Sen., Brit.
* Mactra stultorum, L., Lus., Mog., Brit.
* „ inflata, Bronn.
Corbula revoluta, Broc.
* Diplodonta rotundata, Mont., Can., Brit.
* Lucina lactea, Poli, Can., S. Afr.
* „ pecten, Lam., Lus., Can.
* Mesodesma donacilla, Lam., Lus.
* Donax trunculus, L., Lus., Can., W. Afr.
* Venus verrucosa, L., Lus., Can., S. Afr., Brit.
* „ decussata, L., Lus., Sen., Brit.
* Cytherea exoleta, L., Lus., Sen., Brit.
* „ lunata, Lam., Lus., Brit.
* Cardita calyculata, Brug., Lus., Mog., Can., W. Afr.
* Arca Noæ, L., Lus., Can.,
„ tetragona, Poli, Lus., Can., Brit.
* „ barbata, L., Lus., Brit.?
* „ diluvii, Lam., Senegal.
* Pectunculus violescens, Lam.
* Nucula margaritacea, Lam., Lus., Celt.
* Chama gryphoides, L., Lus., Can., S. Afr.
* Modiola discrepans, Lam., Lus., Brit.
* „ Petagnæ, Scacc., Lus.
* „ lithophaga, L.
* Pinna squamosa, L., Mad.
„ [nobilis, L.]
Spondylus aculeatus, Chem.
* Ostrea cristata, Born.
Patella cærulea, L.; P. scutellaris? Lus., Br.
* „ lusitanica, Gm.
* „ tarentina, Lam.
„ [fragilis, Ph.; P. cærulea.]
* Fissurella græca, L., Lus., Can., Brit.
* „ costaria, Desh.
„ rosea, Lam., Lus., Mog.
* Bulla striata, Brug.
* „ truncata, Adams, Can., Brit.
* Eulima polita, L., Lus., Celt.
* Chemnitzia elegantissima, Mont., Lus., Mog., Can., Brit.

* Truncatella truncatula, Drap., Lus., Celt.
* Paludina thermalis, L.
[Rissoa glabrata, v. M.]
* Natica olla, M. de S.
*   ,, millepunctata, Lam., Lus.
* Nerita viridis, L., Lus., Atlant., Madagascar.
Ianthina bicolor, Atl., Lus., Can.
* Haliotis tuberculata, L., Lus., Mog., Can., Brit.
* Tornatella tornatilis, L.
* Trochus crenulatus, Broc., Can., Lus., Br.
*   ,, striatus, L., Lus.,
*   ,, Adansonii, Payr.
*   ,, varius, Gm.
* Cerithium vulgatum, Brug., Lus., Can.
*   ,, mamillatum, Riss.
*   ,, lima, Brug., Can.
* Cerithium perversum, Brug., Can.
* Fasciolaria lignaria.
* Fusus corneus, L., Lus.
  ,, syracusanus, L.
*   ,, rostratus, Oliv., Can.
* Murex trunculus, L., Can.
[Tritonium variegatum, Lam.]
* Ranella lanceolata, Atl.
Dolium galea, L., Can.
* Buccinum variabile, Ph.
*   ,, mutabile, L., Can.
*   ,, gibbosulum, L.
* Mitra nitescens, Lam.
* Marginella clandestina, Broc., Lus.
*   ,, mileacea, L., Lus.
*   ,, [minuta, Ph.]
Cypræa moneta, L., Lus., Can.
  ,, [erosa, L.]

Of these seventy-six species* as many as forty-seven at least have an Atlantic distribution. Many, as will be seen, occur in our Celtic seas, and many range down the African coast as low as Senegal; whilst some few make their appearance in the South African marine fauna. The species marked

* The species included in brackets are such as may be severally objected to on various grounds. L. signifies the specific Linnean name. The other contractions designate localities or provinces—Atlantic, Lusitanian, Celtic, Madeira, Canaries, Mogador, West African, South African.

by an asterisk occur fossil in the Mediterranean area, and form a large proportion [59 to 70] of such as have had a lengthened settlement in the Mediterranean area.

When treating of marine zones or provinces we are too apt to consider them as defined by hard lines, whereas no certain limits can ever be drawn. The northern Atlantic has certain common specific forms, with a great range from north to south; and, in addition, there are the characteristic forms of the province. With respect to these, the difference presented by each successive breadth of sea consists in the numerical decrease of the individuals of a species from the place where it is typical. As we move south we part with northern forms, and *vice versâ*. This process holds good, not only with the shelled Mollusks, but with every other class. The *Octopus vulgaris* has a great range, and is a common form through several provinces—the Boreal, the Celtic, the Lusitanian, and along the whole of the African coast as far as the Cape. But, as a whole, the Cephalopods can be readily referred to zoological zones. This *Octopus*, however, is common along the eastern coasts of Africa, including the Red Sea; and it occurs there, not because at some past time it passed from the Mediterranean, but because this particular form is equally a constituent of the Eastern as of the West African fauna.

The abundance of stony corals is one of the characteristics of the Red Sea. Some few are common to the Mediterranean, such as *Desmophyllum stellaria* and *D. costatum, Cyathina cyathus, Dendrophyllia ramea, Cladocera cæspitosa* ; but all these range continuously on either side of the great African continent. From the abundance of these particular forms in the east as compared with the west, we may refer them or consider them as belonging to the Indo-Pacific, rather than to the Lusitanian region ; and, conversely, the shells of the foregoing list, which are given as common to the Red Sea and the Mediterranean, and which are all essentially Atlantic species, may have reached their remote eastern settlements by doubling the Cape.

The character of a marine province is dependent in all cases on the preponderance of certain peculiar forms. Perhaps no two lines can be chosen which, though not far apart, will yet present, even to the casual observer, so great a difference as the shore of the Eastern Mediterranean and that of the upper end of the Red Sea. A small set of dead shells picked up near Suez contains *Pleurotoma flavidula, Murex crassispina* and *anguliferus, Cerithium vulgatum, nodulosum,* and *tuberculatum, Cypræa turdus ? Nerita exuvia, Monodonta Ægyptiaca* and *pagodus, Turbo chrysostomus, Trochus maculatus, Fusus Nicobaricus* and *distans, Pyrula citrina, Patella laciniosa, Aspergillum vaginiferum, Cytherea pectinata, erycina,* and two others, *Mactra subplicata, Sanguinolaria rugosa, Spondylus costatus, multilamellatus ? Vulsella lingulata, Arca fusca,* Brug.,

Chama cristella? and a *Lima* distinct from *squamosa*. Only one of these, the wide-spread *Cerithium vulgatum*, is to be found in the Mediterranean; the rest are some of those stranger forms,—the spoils of eastern seas, which attract the attention in collections of such objects.

Had there been a free passage from the Mediterranean to the Red Sea at any time near the present the difference in their respective faunas could hardly have been as great as it is; and whether such communication existed at any former (pliocene) period, must be determined by the amount of agreement between the fossil shells of Italy and Greece with those of the Indo-Pacific region. Without entering on the details of this inquiry, it may be stated generally that the relations of the fossil portion of the Mediterranean fauna are western and Atlantic; and further, that the geologist is unable to give the naturalist any support in his speculations as to a Suez route for the common forms.

CHAPTER XI.

CONCLUSION.

THE map which accompanied the first volume of this series * was on Mercator's projection, and was intended to exhibit the distribution of summer and winter mean temperatures over the European area. When, however, in any branches of natural history the subject of distribution is treated by division of "Zones" or of "Provinces," it is of the utmost importance that relative proportions should be preserved; a projection, such as Mercator's, not only does not suggest any correct views as to the relations of zoological provinces, but may be said to prevent those relations being understood and appreciated.

A map to illustrate the natural history of the European seas has to embrace an area from the whole of the Arctic basin, as far south as Cape Verde, so as to include the South Lusitanian province. The American coasts have to be introduced for the purpose of showing along what extent the Western Atlantic has a molluscous fauna identical

* The Vegetation of Europe.

with ours, and where it is equivalent and representative. Lastly, in an easterly direction it has to extend to the region of the Caspian and Aral Seas.

The map here given is on what is called the "globular projection," and its advantages are, that equal spaces on the sphere are represented by equal spaces on the plane; relative dimensions are preserved; but as the rectangular spaces on the sphere are not represented by like spaces, the *forms* of countries are somewhat distorted. It will be seen that this defect exists to its greatest extent towards the circumference east and west, as in the North Pacific and in the Bay of Bengal; but that for the central portion—that with which we are here more immediately concerned—the meridians and parallels of latitude do not depart much from right-angles. In this map, which is a perspective view of the northern hemisphere, the sphere is represented as it would be seen on the horizon of London at a distance of sixty-eight hundredths of the radius from the surface.

Oceans, seas, lakes, and rivers, mark what are now the depressed portions of the earth's surface; physical arrangements, however, have not always been such as they now are; the past history of the globe presents an almost endless series of changes in the relations of land to water : and as the natural history of our existing seas may enable us to read off the bathymetrical conditions of older sea-beds, so may the arrangement of existing depressions suggest reasons for the forms and relative positions of those older areas of depression which have become effaced, but which will be next described.

The north polar depression is occupied by the Arctic Ocean. The regular form of this basin is well shown in the accompanying map; with the north polar axis as its central point, its shores—both of Asia and Europe—nearly coincide with a circle drawn between 71° and 72° north latitude.

This is the seat of one of our primary zoological provinces, the source of so many forms which at a definite time spread southerly; the great relative magnitude of this province is well seen in the accompanying map.

Physical features having an east and west direction can be traced through other portions of the northern hemisphere. There is a broad zone of depression between the 34th and 47th parallels of north latitude, extending from 10° west to 90° east of the meridian of London; along this are found, in continuous series, the Mediterranean, the Black Sea, the Caspian, the Aral Sea, and the lake system of Central Asia. There are several minor areas of depression running parallel or concentrically with these lines, such as that of 59° north latitude, along which lies the Gulf of Finland, and which is continued from Stockholm eastwards through the Swedish lakes. The north coast of Spain, dependent on the elevation of the Pyrenean range, runs due east and west for upwards of 350 miles. The whole physical framework of Asia and Africa is projected on lines, which run east and west, or on parallels of latitude.

A like arrangement prevails over the whole of the great American continent, such is the valley of the Amazon; these lines are beyond the limits of our present inquiry, and it may suffice to notice that the depression of the North American lake system lies between the same parallels as does that of Asia.

It may be stated generally that throughout the whole of the northern hemisphere there are certain linear areas of depression which are concentric, or which run parallel to one another, at right angles to the earth's polar axis—they conform to lines of equal curvature of the earth's polar compression. These lines of depression have produced our inland seas and lakes, and determined the river systems connected with them; the original formation of these depressions often dates back to periods of considerable antiquity in the earth's history. By means of the testaceous remains of the animals which have tenanted the waters of these depressions, the dates of their formation, and subsequent modifications of extent and form, can be definitely ascertained.

These long linear areas have not been produced at once, and though this subject may seem to belong rather to the geological history of the

European area, it may serve to throw some light on the real nature of the changes which have gone before, if we select one instance and illustration of the manner in which the members of an existing fauna become the evidences of the date of physical change.

The Black Sea fauna, as has been seen, is an extension of that of the Mediterranean; along the whole of the north coast of the Black Sea, from Bessarabia, across to the north-west of the Crimea, and about Kertch, the deposits now forming, and which contain the remains of the existing Black Sea testacea, overlie consolidated beds, which were accumulated beneath the waters of the great Aralo-Caspian Sea (p. 209). The waters of this vast area were fresher than are those of the Aral Sea now, and the dead shells of the present beaches of the Black Sea are in striking contrast with the fresh-water forms which occur in the cliffs—such as *Paludina, Limnæa, Neritina, Melanopsis.*

This fresh-water fauna is of great antiquity, and though vast physical changes have occurred since it first made its appearance in this area, it is not as yet extinct; at least, not wholly so: the fossil *Neritinæ* and *Driessenæ* are identical with those now found in the Danube and the Don; and the aberrant forms of *Cardium,* such as *C. plicatum, C. coloratum,* and *C. pseudo-cardum,* still linger about the mouths of the rivers which now flow into the

Black Sea (the Dnjestr and the Don), as they before did into the Aralo-Caspian.

Physical changes have brought the marine fauna of the Mediterranean in superposition on that of the older Aralo-Caspian basin, and the relative ages of the two are distinctly marked. Should changes again happen—should the amount of evaporation be diminished over this area, or should the escape of the surplus water be arrested by some physical disturbance, which should close the Bosphorus— the fresh waters would again accumulate in the great Aralo-Caspian basin; the fauna of the Black Sea would be gradually extinguished, and the remnant of the older fauna, from the Danube, the Dnjestr, and the Don, would again repeople this great inland sea.

The Black Sea depression has been produced since a large portion of country to the north of it was in the condition of the Aralo-Caspian sea-bed ; and the Black Sea must have received its Testacea after the North Atlantic fauna had extended itself into the Mediterranean (Pliocene period of geologists). The Aralo-Caspian fauna is the oldest with which we are acquainted in connection with the European area.

It will have been seen that the several " Provinces" of our European seas do not admit of limitation by definite lines ; a system of colours graduating from one to another much more truly represents the relations of these assemblages.

The great "Arctic" fauna is here indicated by light blue; a darker tint has been taken for that modification of the Arctic fauna which has been described as "Boreal."

On the south—the West African marine province —from the Senegal river, as far as 25° north latitude, is coloured orange, which, about the Canary islands, passes into yellow. This is the furthest extension which can be given to the marine fauna which has been described as "South Lusitanian." In a northerly direction the Lusitanian province reaches to the Channel Islands, and includes every portion of coast from the Azores to the Black Sea.

The "Celtic" area is coloured green—the blending of the colours of the provinces on either side, as the fauna is composed of the commingling of the forms of those provinces.

The "Boreal" outliers along the west of our Celtic province are indicated by their appropriate dark blue, and the "Celtic" outlier of Vigo Bay (p. 105) by green.

The Caspian province, with its outliers at the mouths of the Danube and the Don, are distinguished by sienna.

The line of "Floating Weed" in the Central Atlantic has been laid down from the small map to the Memoir in which Ed. Forbes first made known those ingenious speculations which have been given succinctly in this volume (p. 110). Ed. Forbes took the position of this "weed" bank from the Physical Atlas of Berghaus; but there are reasons for supposing that it is not quite correctly laid down in that work, particularly in the southern portion.

The configuration of the bed of the Atlantic, and some other considerations, suggest a somewhat different form for that old land which is supposed to have stretched away from the "Old World" into the Atlantic.

A dotted line from Newfoundland to Cape Farewell, (the extreme southern point of Greenland,) thence across the Atlantic by Iceland and the Faroe Archipelago, represents, conjecturally, the northern limitation of the Atlantic at the time when it did not communicate with the Arctic basin.

The northern coast of the connecting land between the Old World and the New may be supposed to have extended continuously from Nordland to North Greenland. This land did not connect itself on the south with the group of the British islands, but passed somewhere to the north, leaving a communication from the Atlantic into the German Ocean, which, at the period of the fauna of the Crag deposits, held to the North Atlantic a like relation of "inland sea" that Hudson's Bay now does on the American continent.

At this period there was no passage from the German Ocean into the English Channel.

The curves of equal winter and summer temperatures are distinguished in the engraving, and three of each set are coloured; the numbers affixed indicate degrees of Fahrenheit's scale.

Such is our brief "outline" of the Zoology of the European Seas; it must be considered as an attempt to present only a general view of the local character, mutual relations, and distribution of the forms of life which tenant the North Atlantic.

The professed Naturalist may perhaps deem our volume to be disproportionate to its subject—and it will disappoint such as may look into it for a repertory of all the forms of our European Seas; but it was not to supply such a want as this that the volume was designed. Viewed as a first attempt, the plan and method of treatment adopted in the earlier portion of the work seem amply sufficient: no one knew better than the late Ed. Forbes that "in the great and wide sea are things creeping *innumerable*, both great and small," and no one better than he could have treated of them fully, had he been so disposed; but that was not his object here.

In Ed. Forbes the Natural History of the "world of waters" experienced its greatest loss: there were his higher investigations and indirectly his personal influence: to associate with him was to feel attracted and interested in his studies, and these "outlines" were undertaken for the purpose of kindling and keeping alive a taste for a branch of knowledge which he felt had a high educational value, and which, in its applications, is, in the end, to unveil the mystery which yet hangs over the early history of our globe.

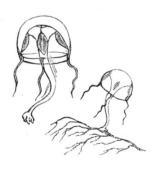

FAUNA OF THE ARCTIC PROVINCE.

Balæna mysticetus, Greenland whale, 36.
Balænoptera boops, Finner, 37.
,, musculus, Rorqual, Spitzbergen, 38.
,, physalis, Razor-back ,, 38.
,, rostrata, ,, 38.
Delphinus leucos, White Whale, Bjeluga, 38.
Monodon monoceros, Narwhal, 38.
Phoca leporina, Nova Zembla, 48.
,, barbata ,, 48.
,, Groenlandica, Harp seal, 48.
,, hispida, 48.
Trichecus rosmarus, Walrus, 48.

Gadus minutus, Capelan, 40, 52.
,, sarda, Nova Zembla, 48.
Liparis vulgaris, 39, 48.
Læmargus borealis, Greenland Shark, 40, 52.
Pollachius virens, Sei, green cod, 39.

Acmæa rubella, 53.
,, testudinalis, 56.
Buccinum glaciale, Spitzbergen, 40.
*Cancellaria viridula, 53, 54. A.
*Fusus islandicus, 49.
Lacuna labiosa, 53, Nordland.
,, frigida, 53.
Littorina rudis, var., Groenlandica, 56.
,, retusa, 54.
Lamellaria prodita, 53, Nordland.

ARCTIC PROVINCE.

Margarita cinerea, 53, 54, 56. A.
Natica aperta, 53. A.
*   „   clausa, 49, 53, 54. A.
*   „   helicoides, 49.
*Purpura lapillus, 49, 56.
Rissoa interrupta, 57, Finmark.
Skenia planorbis, 56.
Scalaria Groenlandica, 54. A.
Scissurella angulata, 53.
Trochus cinerarius, 57, Finmark.
Trophon Gunneri, 53.
      „   harpularum, 54. A.
*   „   scalariforme, 40.
Limacina arctica, 41.
Clio borealis, 41, 53.
*Terebratula psittacea, 40.
      „   septigera, 53, 54, Nordland. A.
*Astarte elliptica, 49.
*   „   corrugata, 49. A.
Hiatella rugosa, 40, Spitzbergen.
*Mitylus edulis, 56.
*Modiola modiolus.
Modiolaria lævigata, 53, 54. A.
Mactra ponderosa, 53, 54. A.
*Mya arenaria, 49.
*   „   truncata, 49, Spitzbergen.
*Panopæa Norvegica, 49
Pecten Groenlandicus, 53, 54. A.
      „   imbrifer, 53. A.
*   „   islandicus, 49.
*Saxicava rugosa, 49.
Ascidia gelatinosa, 40, Spitzbergen.
      „   rustica, 40.
Synoicum turgens, 40.

Brissus lyrifer, 51, Greenland.
Comatula, 47.

*Ctenodiscus polaris, 47.
*Pteraster militaris, 47.
*Ophiocoma acetica, 47.
*Ophioscolex glacialis, 47.
*Ophiocantha spinulosa, 47.
*Ophiolepis Sundevallii, 47.
Beroe cucumis, 45.
Cydippe Flemingii, 45.
Mnemia, 45.

FAUNA OF THE BOREAL PROVINCE.

Phoca barbata, 77.
„ Groenlandica, 74.
„ leporina, 74.
Balænoptera boops, 64.
Delphinus melas, 64.

Anarrhichas lupus, Sea-cat, 74, 104.
*Brosmius vulgaris, 65, Tusk.
Coregonus silus, 65.
*Chimæra monstrosa, 65, 77, King of the Sea.
Clupea harengus, 66.
Cyclopterus lumpus, 74, 104.
Gadus merlangus, 65, 66, 104, Coal-fish, 15-50 fms.
„ merlucius, 65, 66, Hake, 15 fms.
„ morrhua, 66, true Cod, 50 fathoms.
Gymnetrus arcticus, 74, Vaagmer.
*Macrurus Norvegicus, 65.
*Lota abyssorum, 65, 200 fms.
„ molva, 66, 104, Ling, deep open sea.
Pollachius virens, 66, shallow water.
Pleuronectes hippoglossus, 74, Halibut.
*Sebastes Norvegicus, 65, Red-fish, 100 fms.

Spinax niger, 65.

Acmæa testudinaria, 68.
Margarita undulata, 75, below tide line.
Patella pellucida, 69.
„ vulgata, 68.
Purpura lapillus, 68.
Tichrotopis borealis, 75, below tide line.
Æolidia papillosa, 68.
Cuviera squamata, 73.
Chiton marmoreus, 73.
Astarte elliptica, 73.
Cemoria noachina, 73.
Modiola modiolus, 68.
Mya truncata, 74.
Nucula tenuis, 73.
Syndosmya intermedia, 73.
Venus islandica, 74.
Ciona intestinalis, 68.

Echinodermata, 70, 72.
Astrophyton scutatum, 76, Medusa's head, 100 fms.
Brissus fragilis, 72, 100 fms.
„ lyrifer, 73, Christiana, Bute, 10–15 fms.
Cidaris papillata, 72, Norway, Zetland.
Echinus neglectus, 73, Norway, Zetland.
Goniaster equestris, 69, North of Scotland.
„ granulatus, 73.
„ Norvegicus, 76.
Cucumaria frondosa, 75, Great Sea Cucumber, Shetland
Holothuria elegans, 73.
Circe rosea, 76.
Lizzia octopunctata, 76.
Thaumantias pilosella, 76.
Steenstrupia rubra, 76.
Actinia coriacea, 68.
„ mesembryanthemum (?), 68.

*Alcyonium arboreum, 70.
Anthea cereus, 73, 80 fms.
*Oculina ramea, 70.
*   „   prolifera, 73, deep water.
Tethya cranium, 73, 77.

FAUNA OF THE CELTIC PROVINCE.

Accipenser sturio, Sturgeon, 89.
Ammodytes lancea, 103.
Belone vulgaris, Gar-pike, 88.
Blennius vulgaris, Gunnel, 87, 89.
    „   viviparus, 87, 89.
Cottus gobio, Bull-head, 89.
    „   scorpius, 87.
Clupea harengus, 88.
    „   „   var. membras, 88, Balt.
    „   „   var. Cimbrica, 88, Balt.
    „   sprattus, 88, Balt.
Gobius, 89. C.
Gadus callarius, 89, Balt.
Petromyzon marinus, 89.
Platessa plesus, Fluke, 87.
Pleuronectes limandus, 89, Balt.
    „   maximus, 89, Balt.
    „   platessa, 89, Balt.
Trigla cuculus, 103. Bl. Sea, 204.
Spinachea vulgaris, 87.

Aporrhais, 88.
Akera bullata, 87.
Buccinum undatum, 88.
Fusus antiquus, 88.

Littorina rudis, 89, 94, 243, Balt. Bl. Sea, 202.
„   patula, 94.
„   saxatilis, 94.
„   neritoides, 94, 96, 243. Bl. Sea, 202.
„   littorea, 95, 243.
Nassa reticulata, 87.
„   neritea, 92.
Patella vulgata, 94, 244, 265.
„   pellucida, 91, 97, Channel Islands.
Purpura lapillus, 92, 95, 243.
„   hæmastoma, 92, Channel Islands.
Rissoa ulvæ, 89, 244, Balt.
Trochus cinerarius, 96, 243. Bl. Sea, 202.
„   crassus, 96.
„   umbilicatus, 96, 244. Bl. Sea, 202.
„   Laugieri, 93, Channel Islands.
Triton nodiferum, 93,   „   „
„   cutaceum, 93,   „   „
Arca barbata, 93,   „   „
Cardium edule, 89, Balt. Bl. Sea, 202.
Corbula nucleus, 87.
Cyprina islandica, 88.
Donax anatinum, 89, Balt.
Donacilla Lamarckii, 93, Finisterre.
Hiatella arctica, 88.
Leda rostrata, 88.
Lima squamata, 93, Channel Islands.
Limea Sarsii, 101.
Mya arenaria, 89, Balt.
Mytilus edulis, 87, 94, Balt.
Tellina tenuis, 89, Balt. Bl. Sea, 202.
„   solidula, 89, Balt.
„   „   var. Baltica, 90.
Ascidia intestinalis, 87.
Echinus sphæra, 87.
Actinea mesambryanthemum, 96.

Cyathina Smithii, 142.
Desmophyllum Stokesii, 142.
Sphenotrochus Andrewianus, 142.

FAUNA OF THE LUSITANIAN, MEDITERRANEAN, AND BLACK SEA PROVINCE.

Balænoptera musculus, 199, Med. Rorqual, M., L., C.
Delphinus delphis, 198. M., L.
„ tursio, 199. M., Atl.
Pelagus monachus. Monk. 199. E. Med.
Phoca vitulina, 199. M., Boreal.
Sphargis coriacea, 198, Med., W. Af.
Testudo caretta, 198, Med., W. Af.

Accipenser sturio, 207, Bl. Sea.
Atherina presbyter, 122. C., 197, M.
Belone vulgaris, 204, Gar-pike, M., Bl. Sea, Celt.
Beryx decadactylus, 192, Can.
„ splendens, 122. C.
Box salpa, 122. C.
„ vulgaris, 123. C.
Caranx analis, 192. Can.
„ Cuvieri, 123. C.
Chrysophrys cærulosticta, 193. M., Can., W. Af.
Clupea Madeiriensis, 121. C.
Coryphæna equisetes, 192. Can.
Crenilabrus caninus, 121. C.
Diodon reticulatus, 119. C.
Julis Mediterranea, 197. M.
Lampris lauta, 122. C.
Lichia glaucos, 123 C., 193, M.
Mullus barbatus, 197. M.
„ surmuletus, Red Mullet, 123. C.
Oblada melanura, 122. C.

Pagrus vulgaris, 123. C.
Phycis Mediterraneus, 123. C.
Polyprion cernium, 123. C.
Pimelepterus incisor, 192. Can.
Priancanthus boops, 192. Can.
Pristopoma ronchus, 193. M., Can., W. Af.
Prometheus Atlanticus, 122. C.
Sargus cervinus, 193. M., Can., W. Af.
  ,, Rondeletii, 122. C.
Scarus creticus, 196. M.
  ,, mutabilis, 123. C.
Scomber scombrus, 121. C.
Scorpæna scrofa, 121. C.
Sebastes Kuhlii, 121. C.
Serranus cabrilla, 123. C.
Smaris Royeri, 123. C.
Sphyræna vulgaris, 122. C.
Tetrodon marmoratus, 119. C.
Thynnus pelamys, Bonito, 204. M., Bl. Sea.
  ,, vulgaris, Tunny, 204. M., Bl. Sea.
Torpedo narke, 197. M.
Trigon pistanaca, 208. M., Bl. S., Azof, Casp.
Umbrina vulgaris, 204. Bl. Sea.
Xiphias gladius, 206. M., Bl.
Zeus Faber, John Dory, 123. C.

Cephalopoda, 161-165.

Acmæa virginia, 117. L., 178, W. M.
Auricula myosotis, 180. 2-3 fms.
Buccinum gibbosulum, 275. M., Red Sea.
  ,, modestum, 116. L.
  ,, mutabile, 275. M., Can., Red Sea.
  ,, variabile, 275. L., M., Red Sea.
Bulla striata, 202. M., Red Sea, Bl. Sea.
  ,, truncata, 274. C., L., M., Can., Red Sea.
Calyptræa sinensis, 168. M., 202, Bl. Sea.

Cassidaria depressa, 176. E. M.
„ Tyrrhena, 168. M. 176. E. M.
Cassis saburon? 115. L., 168, M.
Cerithium adversum, 202. Bl. Sea.
„ fuscatum, 180.
„ lima, 182, 4, 6. L., M., Red Sea, Can., 2-110 fms.
„ mamillatum, 275. M., Red Sea.
„ perversum, 275. M., Can., Red Sea.
„ vulgatum, 115, 182. L., M., Red Sea, 2-10 fms. 202, Bl. Sea.
Chemnitzia elegantissima, 274. C., L., M., Can., Red Sea.
Chiton cajetanus, 108. L., 179, M.
„ fascicularis, 179. M.
„ fulvus, 117. L.
„ siculus, 169. M., 170, M.
Columbella rustica, 118. L., 179, Marg., 202, Bl. Sea.
Conus Mediterraneus, 115. L., 167, M., 179, 202, Bl. Sea.
Crepidula unguiformis, 168. M.
Cymba olla, 115, 118, 171. L., 178, W. M.
Cypræa erosa, 275. M., Red Sea.
„ moneta, 275. L., M., Can., Red Sea.
„ spurca, 179, 2 fms.
Dentalium novem costatum, 184, 35-55 fms.
„ quinqueangulare, 186, 100 fms.
Dolium galea. L. 176, 177. E. M., Can., Red Sea.
Emarginula elongata, 184, 55-80 fms.
„ Huzardi, 180, 2-3 fms.
Eulima polita, 274. C., L., M., Red Sea.
Fasciolaria lignaria, 275. M., Red Sea.
„ Tarentina, 179, 2 fms.
Fissurella costarea, 274. M., Red Sea.
„ græca, 274. C., L., M., Can., Red Sea.
„ rosea, 274. C., L., Mog., Red Sea.
Fusus contrarius, 274, 109, 113.
„ corneus, 169. M., L., Red Sea.
„ lignarius, 179, 2 fms.

Fusus muricatus, 185, 80-100.
,, rostratus, 275. Can., M., Red Sea.
,, Syracusanus, 275. L., Red Sea.
Haliotis lamellosus, 179, Marg.
,, tuberculata, 171. L., M., Red Sea, Can.
Jeffreysia cærulescens, 181.
,, diaphana, 170.
Littorina neritoides, 117. L., 202, Bl. Sea.
,, petræa, 179. M. 181.
,, rudis, 202, Bl. Sea.
Marginella clandestina, 182. 275. C., M., R. Sea, 2-10 fms.
,, mileacea, 275. C., M., Red Sea.
,, minuta, 275. M., Red Sea.
Mesalia striata, 178. W. M.
,, sulcata, 178. W. M.
Mitra obsoleta, 182, 2-10 fms.
Murex Brandaris, 115. L., 167, M.
,, corallinus, 115. L.
,, cristatus, 176. E. M.
,, Edwardsii, 115. L.
,, erinaceus, 202, Bl. Sea.
,, trunculus, 115. L., 167-9, 172. M., Can., Red Sea.
Nassa Ascanias, 202, Bl. Sea.
,, gibbosula, 180, 2-8 fms.
,, neritea, 180, 2-3 fms.
,, reticulata, 202, Bl. Sea.
Natica Guilleminii, 115. L.
,, intricata, 115. L., 178, W. M.
,, olla, 182, 2-10 fms.
Parthenia fasciata, 186, 100 fms.
,, ventricosa, 186, 100 fms.
Patella Bonnardi, 179, Marg.
,, cœrulea, 274. C., L., M., Red Sea.
,, ferruginea, 202, Bl. Sea.
,, fragilis, 274. M., Red Sea.
,, Lusitanica, 274. L., M., Red Sea.
,, pellucida, 108, 170. N. L., W. of Af.

Patella scutellaris, 169. M., 179, Marg.
,, tarentina, 202, Bl. Sea, M., Red Sea.
Pedicularia sicula, 176. E. M.
Phasianella intermedia, 115. L.
,, pulla, 202, Bl. Sea.
Pleurotoma costulatum, 202. M., Bl. Sea.
,, Maravignæ, 108, 184. L., M., 55-80 fms.
Pollia maculosa, 179.
Purpura hæmastoma, 115. L.
,, lapillus, 170, Vigo.
Ranella lanceolata, 275. L., M., Red Sea.
Ringicula auriculata, 108, 116. L., 171, M.
Rissoa oblonga, 182, 2-10 fms.
,, pulcherrima, 170.
,, reticulata, 185-6, 80-100 fms.
,, ventricosa, 182, 2-10 fms.
Scalaria hellenica, 186, 100 fms.
Siphonaria concinna, 118. L., 178, W. M.
Solarium luteum, 171.
,, stramineum, 171.
Triton corrugatum, 115. L.
,, cutaceum, 115. L.
,, variegatum, 108, 119. L., M., Red Sea.
Trochus Adansonii, 202, M., Red Sea.
,, canaliculatus, 115. L.
,, cinerarius, 109, 170. L., M., W. of Af., 202, Bl. Sea.
,, crenulatus, 182. C., L., M., Can., Red Sea, 2-10 fms.
,, divaricatus, 202, Bl. Sea.
,, exiguus, 202, Bl. Sea.
,, fragarioides, 166. M., 202, Bl. Sea.
,, Laugieri, 108, 115. L., M.
,, Lyciacus, 179. E. M.
,, millegranus.
,, Sprattii, 176, 182. E. M., 2-10 fms.
,, striatus, 275. C., M., Red Sea.

Trochus tumidus, 109, 170. L., M.
„ umbilicatus, 118. L., 202, Bl. Sea.
„ varius, 275. M., Red Sea.
„ villicus, 202. M., Bl. Sea.
„ ziziphinus.
Tornatella tornatilis, 275. L., M., Red Sea.
Turbo cuneatus, 55–80 fms.
„ rugosus, 115. L.
„ sanguineus, 184–5, 80–100 fms.
Turritella communis, 168. M.
„ incrassata, 114.
„ sulcata, 118. L.
„ tricostata, 182, 35–55 fms.
Truncatella truncatula, 180, 181, 2–3 fms., 202, Bl. Sea.
Umbrella Mediterranea, 176. E. M.
Velutina lævigata, 108, 170. N. L.

Amphidesma sicula, 180. M., 2 fms., sand.
Anomia ephippium.
Arca barbata, 179, 274. L., M., Red Sea.
„ diluvii, 274. Sen., Red Sea.
„ imbricata, 186. 230 fms.
„ lactea, 186, 2–100 fms.
„ Noæ, 274. L., M., Can.
„ scabra.
„ tetragona, 274. C., L., M., Can., Red Sea.
Astarte sulcata, 178. W. M.
„ triangularis, 178. W. M.
Bornia corbuloides, 115. L.
Cardita aculeata, 185, 80–100 fms.
„ calyculata, 179. L., M., Mog., Can., W. Af.
„ trapezia, 116. L.
Cardium coloratum, 203, Bl. Sea.
„ edule, 180. M., 202, Bl. Sea.
„ exiguum, 181, 202, Bl. Sea, 2–10 fms. C.
„ papillosum, 181–4, 2–10 fms. C.
„ plicatum, 203, Bl. Sea.

Cardium rusticum, 115. L.
Chama gryphoides, 168. L., M., Can., S. Af.
Clavagella angulata, 176. E. M.
„ balanorum, 176. E. M.
„ Melitensis, 176. E. M.
Corbula revoluta, 274. M., Red Sea.
Cytherea exoleta, 274. C., L., Sen., Red Sea.
„ lincta, 274. C., L., Red Sea.
„ Venetiana, 116. L.
Diplodonta rotundata, 274. C., L., M., Can., Red Sea.
Donax trunculus, 180. L., M., Can., W. Af., Red Sea, 2-3 fms.
Ervilia castanea, 115, 178. L., W. M.
Erycina ovata, 202. M., Bl. Sea.
Kellia corbuloides, 180, 2-3 fms.
Leda emarginata, 116. L.
Ligula profundissima, 186, 100 fms.
Lima crassa, 186, 100 fms.
„ elongata, 185, 80-100 fms.
„ squamosa, 179.
Lithodomus caudigerus, 118. L.
„ dactylus, 168. M.
„ lithophagus, 118. L., 179, M.
Lucina Desmarestii, 180, 2 fms., sand.
„ digitalis, 116. L.
„ divaricata, 116. L.
„ lactea, 180. L., M., 202, Bl. Sea, Can., S. Af., Red Sea, 2 fms., mud.
„ pecten, 180. L., Can., Red Sea, 23 fms.
Lutraria elliptica, 178. W. M.
Mactra helvacea, 115. L.
„ inflata, 274. L., M., Red Sea.
„ rugosa, 171.
„ stultorum, 180. C., L., M., W. Af., Red Sea, 23 fms.
„ subtruncata, 100, 170. L., M., Vigo.
„ triangula, 202. W. M., Bl. Sea.

Mesodesma donacilla, 170. L., M., Red Sea, 2 fms., sand.
Modiola discrepans, 274. C., L., M., Red Sea.
,, lithophaga, 274. M., Red Sea.
,, Petagnæ, 274. L., M., Red Sea.
Mytilus afer, 116–18. L.
,, minimus, 115. L., 169, M., 202, Bl. Sea.
Nucula Ægeensis, 186, 100 fms.
,, emarginata, 183. M., 20-35-55 fms.
,, margaritacea, 181. C., L., M., Red Sea, 2-10 fms.
,, striata, 184, 25-55-80 fms.
Ostrea Adriatica, 212, Bl. Sea.
,, cristata, 274. M., Red Sea.
,, edulis, 117. L.
Panopæa Aldrovandi, 115. L.
Pecten Hoskinsii, 186, 100 fms.
,, hyalinus, 180. M., 2-10 fms.
,, Jacobæus, 176. E. M.
,, maximus, 117. L.
,, opercularis, 183. M., 35-55 fms.
,, polymorphus, 116. L., 181, M., 2-10 fms.
,, similis, 186, 50-80-100 fms.
,, tigerinus, 180, Vigo.
Pectenculus violescens, 274. M., Red Sea.
Petricola lithophaga, 115. L.
Pholas candida, 202, Br., M., Bl. Sea.
Pinna squamosa, 274. L., Med., Red Sea.
Psammobia rugosa, 118. L.
Solecurtus strigillatus, 115. L., 179, M., 2 fms., sand.
Solemya Meditterranea, 108. M., 2 fms., sand.
Solen ensis, 202. Br., M., Bl. Sea.
,, legumen, 274. C., L., M., Red Sea.
,, vagina, 274, Celt., L., M., Red Sea.
Spondylus aculeatus, 274. M., Red Sea.
,, gaderopus, 168–79. M.
Tapes pallustra, 109, Vigo.
Terebratula appressa, 185, 80-100 fms.
,, cuneata, 185, 20-35 fms.

Terebratula detruncata, 184, 20-35 fms.
,, lunifera, 185, 80-100 fms.
,, seminula, 184, 35-55 fms.
,, truncata, 185, 55-80 fms.
,, vitrea, 185, 80-100 fms.
Tellina balaustina, 170.
,, carnaria, 202. Br., M., Bl. Sea.
,, costæ, 116. L.
,, distorta, 116. L., 181, M., 2-18 fms.
,, donacina, 180. M., 2-10 fms.
,, tenuis, 202. Br., M., Bl. Sea.
Thecidea Mediterranea, 176. E. M.
Venus aurea. 202. Br., M., Bl. Sea.
,, dysera, 202. Br., M. Bl. Sea.
,, gallina, 202. Br., M.. Bl. Sea.
,, ovata, 184, 50-80-100 fms.
,, striatula, 117. L., 178, W. M.
,, verrucosa, 274. C., L., M., Can., W. Af., Red Sea.
Venerupis decussata, 176, 180. E. M.
,, irus, 202. Br., M., Bl. Sea.

Amouroucium argus. 158. Br., M.
Ascidia arachnoidea, 158. Br., M.
,, mentula, 158. Br., M.
,, microcosmus, 159.
,, rustica, 159. E. M.
Botryllus papillosa, 158. Adr.
,, polycyclus, 158. Ar., M.

Pteropods, 159.
Cleodera pyramidata, 160. L., M.
,, subula, 160. L., M.
Hyalæa tridentata, 160. L. M.
,, trispinosa, 160. L., M.

Bryozoa, 145.

Crustacea, 153-7.
  Acanthonyx lunulata. L.
  Amathia Rouxii. L.
  Athanas nitescens. C.
  Calappa granulata. L.
  Cancer pagurus. C., Max.
  Carcinus mænas. C., Max.
  Dorippe lanata. L.
  Grapsus Messor. Can.
  „ strigosus. Can.
  Hemola hispida. L.
  „ spinifrons. L.
  Herbstia nodosa. L.
  Hyas coarctatus. C.
  Inachus dorynchus. L., M., Can.
  Labretes elegans. L.
  Leptopodia sagittaria. Can., West Ind.
  Lupa hastata. L.
  Lissa. L., Gaulteri.
  Mithrax dichotoma. L.
  Nephrops Norvegicus. C., M.
  Ocypode ippeus. L.
  Pandalus annulicornis. C.
  Plagusia clavimana. Can. and W. Af.
  Polybius Henslowii. C., W., and S.
  Portunus puber. C., Max.
  Scyllarus latus. L.
  Squilla Desmarestii. L.
  „ mantis. L.

Echinoderms, 148–153.
  Amphiura neglecta. M.
  Asterias aurantiaca. M., L., Br.
  „ tenuispina. M., L.
  Asterica gibbosa (minuta). M., L., B.
  Astrophyton scutatum. M., L., Boreal.
  Astropyga, Can.

Brissus scillæ.
" ventricosus. M., L., Can.
Cidaris hystrix. M., L.
" imperialis, Med., Can.
Comatula rosacea. M., L., Brit.
Cucumaria pentacte. M., Boreal.
Echinocyamus pusillus. M., L., Br.
Echinus esculentus. M., L.
" lividus. M., L., Br.
" melo. M., L.
" monile. L., M.
" Sardicus. M.
Ophidiaster granifera. M., L., Can.
" ophidiana. M., L., Can.
Ophiocoma scolopendroides. M., L., Br.
Ophiothrix rosula. M.
Ophiura abyssicola. M.
" albida.
" lacertosa. M., L., Br.
" texturata. M.
Palmipes membranaceus. M., L., Br.
Spatangus purpureus. M., L., B.
Stellonia glacialis. M. Boreal.
" tenuispina, St. Lusit., M.
Syrinx nudus. M., Boreal.
Uraster glacialis.

Medusæ, 143.

Astroides calycularis, 142. M.
Balanophyllia Italica, 142. L., M.
" verrucaria, 142. L., M.
Cladocera astræaria, 142. M.
" cespitosa, 142. M.
" stellaria, 142. M.
Cœnocyathus anthophyllitis, 142. M.
" Corsicus, 142. M.

Cyathina cyathus, 142. L., M.
" pseudoturbinolia, 142. L., M.
Dendrophyllia cornigera, 142. L., M.
" ramea, 142. L., M.
Desmophyllum cristagalli, 142. L.
" stellaria, 142. L., M.
Sphenotrochus millitianus, 143. L.

Actinia mesembryanthemum, 141, Med., Atl., Red Sea.
" tapetum, 141, Med., Atl., Red Sea.
Antipathes subpinnata, 140. L., M., Can.
Corallium rubrum, 139. M., Red Sea.
Funicularia quadrangularis, 141, Med., Atl.
Gorgonia ceratophyta, 140. M., Atl.
" coralloides, 140. M., Atl.
" placomus, 140. M., Atl.
" tuberculata, 140. M.
Isis elongata, 139. M., Ind. Oc.
Lobularia palmata, 138. M., L.
Pennatula phosphorea, 140. M., Atl.
" setacea, 141, Med., Can.

Tethya lyncurium, 134. M., Brit.

Foraminifera, 134.

Quinqueloculina subrotunda, 136, Brit.
Truncatulina lobata, 135. M., Brit.

In the foregoing lists the Provinces are indicated by their initial letters. See page 275.

HISTORY OF ECOLOGY
An Arno Press Collection

Abbe, Cleveland. **A First Report on the Relations Between Climates and Crops.** 1905

Adams, Charles C. **Guide to the Study of Animal Ecology.** 1913

American Plant Ecology, 1897-1917. 1977

Browne, Charles A[lbert]. **A Source Book of Agricultural Chemistry.** 1944

Buffon, [Georges-Louis Leclerc]. **Selections from Natural History, General and Particular, 1780-1785.** Two volumes. 1977

Chapman, Royal N. **Animal Ecology.** 1931

Clements, Frederic E[dward], John E. Weaver and Herbert C. Hanson. **Plant Competition.** 1929

Clements, Frederic Edward. **Research Methods in Ecology.** 1905

Conard, Henry S. **The Background of Plant Ecology.** 1951

Derham, W[illiam]. **Physico-Theology.** 1716

Drude, Oscar. **Handbuch der Pflanzengeographie.** 1890

Early Marine Ecology. 1977

Ecological Investigations of Stephen Alfred Forbes. 1977

Ecological Phytogeography in the Nineteenth Century. 1977

Ecological Studies on Insect Parasitism. 1977

Espinas, Alfred [Victor]. **Des Sociétés Animales.** 1878

Fernow, B[ernhard] E., M. W. Harrington, Cleveland Abbe and George E. Curtis. **Forest Influences.** 1893

Forbes, Edw[ard] and Robert Godwin-Austen. **The Natural History of the European Seas.** 1859

Forbush, Edward H[owe] and Charles H. Fernald. **The Gypsy Moth.** 1896

Forel, F[rançois] A[lphonse]. **La Faune Profonde Des Lacs Suisses.** 1884

Forel, F[rançois] A[lphonse]. **Handbuch der Seenkunde.** 1901

Henfrey, Arthur. **The Vegetation of Europe, Its Conditions and Causes.** 1852

Herrick, Francis Hobart. **Natural History of the American Lobster.** 1911

History of American Ecology. 1977

Howard, L[eland] O[ssian] and W[illiam] F. Fiske. **The Importation into the United States of the Parasites of the Gipsy Moth and the Brown-Tail Moth.** 1911

Humboldt, Al[exander von] and A[imé] Bonpland. **Essai sur la Géographie des Plantes.** 1807

Johnstone, James. **Conditions of Life in the Sea.** 1908

Judd, Sylvester D. **Birds of a Maryland Farm.** 1902

Kofoid, C[harles] A. **The Plankton of the Illinois River, 1894-1899.** 1903

Leeuwenhoek, Antony van. **The Select Works of Antony van Leeuwenhoek.** 1798-99/1807

Limnology in Wisconsin. 1977

Linnaeus, Carl. **Miscellaneous Tracts Relating to Natural History, Husbandry and Physick.** 1762

Linnaeus, Carl. **Select Dissertations from the Amoenitates Academicae.** 1781

Meyen, F[ranz] J[ulius] F. **Outlines of the Geography of Plants.** 1846

Mills, Harlow B. **A Century of Biological Research.** 1958

Müller, Hermann. **The Fertilisation of Flowers.** 1883

Murray, John. **Selections from *Report on the Scientific Results of the Voyage of H.M.S. Challenger During the Years 1872-76.*** 1895

Murray, John and Laurence Pullar. **Bathymetrical Survey of the Scottish Fresh-Water Lochs.** Volume one. 1910

Packard, A[lpheus] S. **The Cave Fauna of North America.** 1888

Pearl, Raymond. **The Biology of Population Growth.** 1925

Phytopathological Classics of the Eighteenth Century. 1977

Phytopathological Classics of the Nineteenth Century. 1977

Pound, Roscoe and Frederic E. Clements. **The Phytogeography of Nebraska.** 1900

Raunkiaer, Christen. **The Life Forms of Plants and Statistical Plant Geography.** 1934

Ray, John. **The Wisdom of God Manifested in the Works of the Creation.** 1717

Réaumur, René Antoine Ferchault de. **The Natural History of Ants.** 1926

Semper, Karl. **Animal Life As Affected by the Natural Conditions of Existence.** 1881

Shelford, Victor E. **Animal Communities in Temperate America.** 1937

Warming Eug[enius]. **Oecology of Plants.** 1909

Watson, Hewett Cottrell. **Selections from *Cybele Britannica.*** 1847/1859

Whetzel, Herbert Hice. **An Outline of the History of Phytopathology.** 1918

Whittaker, Robert H. **Classification of Natural Communities.** 1962